Bioelectricity and Morphogenesis

The Electric Blueprints of Life

Team Trex

Published by NobleTrex Press

For permissions and other inquiries, write to:
P.O. Box 3132, Framingham, MA 01701, USA

Contents

Introduction

Cut a standard planarian flatworm into three pieces, and something miraculous happens. The head grows a new tail. The tail grows a new head. The middle section, suspended in a moment of biological indecision, grows both a new head and a new tail. Within weeks, you have three perfect, genetically identical worms. This party trick of biology is well known; high school students do it in labs every year. But recently, the script changed.

In a laboratory at Tufts University, researchers manipulated the faint, invisible electrical conversation happening between the cells of a cut worm. They didn't touch the genome. They didn't edit the DNA. They simply altered the voltage gradient-the electrical "pressure"-across the wound. The result was a worm with two heads, one at each end. When these two-headed worms reproduced, their offspring also had two heads, despite their DNA remaining completely normal.

This is not a story about genetic mutation. It is a story about the software of life.

For the better part of a century, we have been living in the age of the Genome. We have been told that DNA is the master blueprint, the unassailable commander of our biological destiny. We cracked the double helix, mapped the human genome, and built CRISPR to edit the code. We assumed that if we just cataloged enough genes, we would understand how a single fertilized egg transforms into the staggering complexity of a human face, a beating heart, or a functioning hand.

But we were missing half the picture. If DNA is the hardware-the hard drive containing the parts list for the proteins that make up our bodies-then bioelectricity is the operating system. It is the invisible architect that tells those proteins where to go, when to build, and when to stop.

This book is about that operating system. It is about the realization that electricity is not just for our brains and our nerves, but for every single cell in our bodies. It is about the discovery that our tissues form a collective intelligence, a "hive mind" of cells that uses electrical signals to store memories of what our anatomy *should* look like-and how we are learning to hack that code to fix birth defects, cure cancer, and perhaps even regrow lost limbs.

To understand why this matters, you have to look at your own hand. How did the cells in your thumb know to become a thumb? Genetically, they are identical to the cells in your pinky. They contain the exact same DNA. Yet, during your development, they behaved differently. They stopped dividing at different times; they arranged themselves into different shapes.

For decades, biologists searched for the "thumb gene," but it doesn't exist. Instead, what exists is a system of positional information. We now know that long before the physical structures of your face or limbs appear in the embryo, a "pre-pattern" creates a ghostly scaffolding made of voltage. As shown by the pioneering work of researchers like Michael Levin, if you can visualize the electrical charges in a developing frog embryo, you will see the "face" light up in the voltage map days before the physical face forms. The electricity draws the map; the cells merely color it in.

This completely upends how we think about the body. We are accustomed to thinking of the body as a machine made of chemical gears. If a gear breaks, we try to fix it with a chemical wrench (a drug). But if the body is actually a computer running a bioelectric program, we have been going about medicine all wrong. We have been trying to fix software bugs by soldering the hard drive.

The implications of this shift are profound. If we can learn to read and write this electrical code, we stop being mechanics and start becoming programmers. Instead of trying to micromanage every cell with drugs, we could simply issue a command: "Build a kidney here." "Repair this spinal cord." "Stop this tumor."

The story of how we arrived here is filled with false starts and forgotten pioneers. As we will explore, the connection between electricity and life goes back to the twitching frog legs in Luigi Galvani's 18th-century studio. But for a long time, the study of bioelectricity was overshadowed by the glamorous rise of molecular genetics. Electricity was relegated to the neuroscientists, who rightly marveled at the speed of the brain but largely ignored the slower, steady electrical hum of the rest of the body.

It turns out that neurons are just the speed demons of the cellular world. They evolved to carry signals fast, for movement and thought. But the machinery they use-ion channels, voltage gates, synapses-is actually ancient. It was invented by evolution billions of years ago, around the time single-celled organisms decided to become multicellular. To work together, cells needed a language. Electricity, with its ability to propagate signals rapidly across distances, was the perfect medium.

Every cell in your body is a tiny battery. It spends a tremendous amount of energy pumping ions-charged particles like potassium and sodium-in and out of its membrane to maintain a specific voltage. This isn't the high-voltage current that runs your toaster; it is a subtle, steady field. When these cells connect to their neighbors, they form a network. Just as the internet allows millions of computers to process information that no single laptop could handle, tissues form a computational network that holds the "memory" of the anatomy.

We see this memory in action in the salamander. You can cut off a salamander's leg, and it will grow back, bone for bone, muscle for muscle. Once the leg is complete, the growth stops. How does it know when it's done? The salamander possesses a bioelectric setpoint-a vision of the "correct" leg-that persists even when the physical leg is gone. The cells consult this blueprint and build until their reality matches the plan. Humans, sadly, lose this ability as we age, but the machinery is still there, dormant, waiting for the right signal to wake it up.

The consequences of this system breaking down are just as revealing as when it works. We tend to view cancer as a genetic disease, a series of unlucky mutations that make cells grow uncontrollably. While mutations initiate the problem, the progression of cancer is largely a failure of communication.

Think of a normal organ as a choir, singing in harmony. A cancer cell is a member of the choir that has stopped listening to the conductor. It essentially goes deaf to the bioelectric signals of its neighbors. It reverts to a unicellular state, treating the rest of the body as a hostile environment. It's not necessarily "evil"; it's just disconnected. Research has now demonstrated that if you can artificially force a tumor cell to reconnect electrically with its neighbors-if you can "open the chat lines" again-it will often normalize. It remembers it is part of a kidney or a lung and stops growing, despite the mutations in its DNA. This offers a breathtaking new avenue for cancer therapy: normalization rather than destruction.

But we are doing more than just observing these phenomena; we are beginning to engineer them. The toolkit of the modern biologist has evolved from the scalpel and the microscope to light-activated ion channels and artificial intelligence. We can now use optogenetics-a technology that uses light to control cells-to "type" commands directly into living tissue.

In one startling experiment, researchers induced an eye to grow on the gut of a tadpole simply by changing the voltage of the gut cells to match the pattern of an eye. The gut cells, which were genetically programmed to digest food, happily switched tasks. They built a complete, functioning eye, complete with lens and retina, and even wired it into the nervous system so the tadpole could see through it. This proves that cells are not hardwired robots; they are versatile agents willing to build whatever the bioelectric commander orders.

This plasticity has led to the creation of entirely new forms of life. In 2020, a team involving researchers from the University of Vermont and Tufts University announced the creation of "Xenobots." These were not robots made of metal and plastic, but living machines designed by a supercomputer and built from frog skin and heart cells. Freed from the constraints of the frog's body plan, these cells organized themselves into new shapes, moving and working together in ways never seen in nature. They are a proof of concept that the "rules" of biology are far more flexible than we imagined. We are entering an era of programmable biology, where we are limited not by the genes we inherit, but by the designs we can compute.

As we peel back the layers of this electric blueprint, we are forced to confront deep philosophical questions about what we are. We are not merely the sum of our genes. We are the result of a dynamic, ongoing computation. Our shape, our health, and our very identity are maintained by a crackling web of communication that never sleeps.

This field, often called "morphoceuticals" or regenerative medicine, suggests that the future of healing won't just be about killing germs or replacing parts. It will be about information. We are moving toward a time when a birth defect could be corrected in the womb not by surgery, but by a brief electrical pulse that reminds the cells where the heart should go. We may see a time when severe injuries are treated with "electroceutical" bandages that impose the voltage map of a healthy limb, guiding the body to rebuild itself.

Of course, with great power comes the inevitable need for caution. The ability to rewrite the body plan touches on the ethics of control, identity, and the definition of natural life. If we can redesign organisms, where do we stop? If we can reboot the body's software, does that change the self? These are not questions for a distant sci-fi future; the technology is already fermenting in petri dishes around the world.

The story I am about to tell you is a synthesis of physics, biology, and computer science. It moves from the microscopic gates on the surface of a cell to the macroscopic symmetry of the whole body. It is a look at life not as a static sculpture, but as a dynamic flame-a structure maintained by the continuous flow of energy and information.

We often look for the secret of life in the permanence of DNA, preserved in amber or extracted from ancient bones. But life is defined by its ability to change, to adapt, to repair, and to become. That spark-that ability to hold a shape against the chaos of the universe-is electric.

So, let us unplug our old assumptions. It is time to look at the invisible forces that sculpt the clay of our existence. It is time to meet the invisible architect.

Chapter 1

The Spark in the Machine

Long before neurons evolved to twitch a muscle or conceive a thought, single-celled ancestors were already humming with electrical potential. This chapter peels back the skin to reveal the pervasive bioelectric layer that powers every cell in the body, revisiting a scientific controversy that began with a dead frog in the 18th century and ended with the realization that we are all made of living batteries.

1.1 The Frog That Jumped Forever

Bologna in the 1780s was the sort of place where a serious person could spend the afternoon dissecting a frog and the evening debating philosophy, and feel no contradiction. Luigi Galvani did both. He was a physician and anatomist, not a showman, and he treated the twitching of animal tissue as a clue to something profound: that living bodies were not merely *chemically* reactive but *electrically* organized.

The famous frog's leg enters the story the way many scientific protagonists do: as a humble object that refuses to behave quietly. A frog's hind limb, prepared in a particular way—with a piece of nerve exposed, muscle intact—could be made to kick when touched by metal.

Sometimes it kicked when the weather turned stormy. Sometimes it kicked when Galvani's assistants happened to touch a brass hook and an iron railing in the right sequence, closing a circuit without meaning to. There is a reason these anecdotes persisted. They have the theatrical simplicity of a magic trick, and yet they also have the unsettling quality of a message: the body is already wired.

Galvani's interpretation was bold but not mystical. He proposed that animals possess their own electricity, a kind of internal charge that nerves and muscles can store and release. He used the language available to him—*animal electricity*—and the idea sounded, to some ears, like a return of vital forces under a new name. But his experiments were not pleas for mystery. They were attempts to make physiology measurable. He was trying to say, with the tools of his time, that the difference between a living nerve and a dead one might be expressed in something like volts.

That claim landed in an intellectual landscape already electrified, in every sense. Europe had been captivated by the ability to produce static sparks, to shock party guests, to make dead limbs twitch with friction machines. Electricity was the fashionable phenomenon that straddled salon entertainment and serious inquiry. Yet the electricity of the laboratory tended to be treated as an external force applied to passive matter. What Galvani suggested was more intimate: electricity as a property *generated and managed* by living tissue.

Not everyone agreed, and the most formidable dissenter was Alessandro Volta. Volta was not a mere contrarian; he was a physicist with a deep feel for instruments and materials. Where Galvani looked at the frog and saw a biological battery, Volta looked at the metals and saw chemistry. Perhaps, Volta argued, the electricity did not originate in the animal at all. Perhaps it arose from contact between different conductors—two dissimilar metals connected by a moist medium. In that view the frog's leg was not a generator; it was a detector, a sensitive electroscope that happened to contract when current passed through it.

The debate between Galvani and Volta is often retold as a clean victory for Volta, because Volta went on to build the voltaic pile: a stack of alternating metals separated by brine-soaked material, producing a steady current. This was a revolutionary device. It was electricity you could turn on and keep on, not a fleeting spark. It helped create electrochemistry and eventually the battery-powered world. In that triumphal arc, Galvani becomes the earnest anatomist who misunderstood his own experiment.

That tidy ending is emotionally satisfying and historically misleading. Volta was right that dissimilar metals can generate current in a moist environment, and his pile demonstrated it. But Galvani was right, too, in a way that could not yet be cleanly measured: living tissues maintain electrical differences across their membranes, and nerves and muscles are exquisitely responsive to electrical currents because they are already electrical systems. Galvani's frog leg, far from being a mere detector, was a piece of excitable tissue whose behavior only makes full sense once one accepts that excitability is an electrical property of cells.

A modern reader might wonder how two brilliant people could look at the same twitch and disagree about its source. The answer is that the phenomenon had more than one cause, and the experimental world of the 1780s made it hard to separate them. Metal contact potentials and biological potentials can both contribute to a current. A frog preparation is damp, salty, full of ions. Metal probes differ in their tendency to give up electrons. The interface between metal and tissue has its own electrochemical complexities. With no voltmeter capable of peering into cells, no clear concept of membrane potential, and no ion channel theory, it was almost inevitable that the debate would become a referendum on what sort of explanation one preferred.

What matters for the present story is not who *won* but what the controversy *did* to biology. It helped create a conceptual border that hardened over the next century and a half: the border that said electricity belongs to nerves and muscles, while everything else in life belongs to chemistry.

At first, this border looked like common sense. Nerves conduct signals quickly; muscles contract dramatically. If electricity is about sparks and shocks, surely it fits there. And as physiology matured, it built an extraordinarily successful science of excitability. The nerve became a cable, the muscle a motor. Electrical metaphors became precise mathematics. By the mid-twentieth century, the work of Hodgkin and Huxley on the squid giant axon could describe an action potential as changing conductances of sodium and potassium, with the membrane behaving like a capacitor. You could write equations that predicted the shape of a spike and then measure that spike on an oscilloscope. It was one of the most satisfying matches between theory and life that biology has ever achieved.

Meanwhile, the rest of biology was constructing its own intellectual empire around molecules. Enzymes, metabolic pathways, DNA, RNA, proteins, gene regulation—a chemical grammar of life that explained heredity, development, and disease with astonishing power. In this world, electricity seemed like a special trick used by specialized tissues, interesting but not central.

The split was never officially declared; no one held a congress and voted. It happened the way many scientific partitions happen: through training, through textbooks, through the kinds of questions that seem respectable to ask. If a graduate student proposed that a frog embryo might pattern its body plan partly through long-range electrical gradients, the idea could sound, for decades, like category confusion. Embryos were the domain of genes and morphogens. Electricity was the domain of neurons.

The tragedy of this division is subtle. It did not prevent progress; in many ways it enabled it. Narrowing the scope of electrical thinking to nerves allowed electrophysiology to become exquisitely detailed. Narrowing the scope of heredity and development to molecules allowed genetics to flourish. The tragedy is that, in exchange for this efficiency, biology quietly lost a unifying language: a way to talk about the body as an electrical system *everywhere*, not just in the brain.

Galvani's frog, then, jumped forever not because it was a curiosity but because it kept returning as a provocation. Each time biology confronted a phenomenon that looked like long-range coordination—a wound that heals with directional growth, an embryo that reliably positions organs, a salamander limb that regrows with correct anatomy—the question resurfaced: is chemistry alone enough to choreograph this, or is there an electrical layer of control humming beneath the molecular bustle?

To see why that question is not romantic nostalgia for an old debate, consider what Galvani was actually doing with his frog leg. He was not merely shocking tissue and marveling at motion. He was probing the link between *signal* and *form*. Muscle contraction is a visible output, a readout of an invisible control system. If a twitch can be triggered by closing a circuit, then the boundary between mechanical movement and electrical state has been breached. That breach is what makes the nervous system possible, but it is also what makes any coordinated tissue behavior possible. A cell that can change its membrane voltage can, in principle, change what it does: divide, move, secrete, differentiate, connect.

Volta's contribution, paradoxically, also points in this direction. The voltaic pile made electricity reliable and continuous. Once you can deliver a steady current, you can begin to ask a different class of questions. Not *can* electricity make tissue twitch? but *how much, how long, with what polarity, with what spatial pattern*? Those are exactly the kinds of questions that modern developmental bioelectricity asks. The frog leg was an early bioassay. Volta gave the world a dial.

The historical irony is that, for a long time, electricity became a tool used *on* biology rather than a property recognized *in* biology. Batteries powered stimulation experiments; they powered early electrophysiological instruments; they powered the metaphors of "current" and "circuit" that helped describe nerve conduction. But the idea that ordinary tissues might maintain instructive electrical patterns—patterns that help tell cells where they are and what they should become—was treated as marginal. Galvani's animal electricity survived in the narrow sense of "nerves are electrical," while the broader implication was quietly shelved.

A branching diagram can make this historical bifurcation visible.

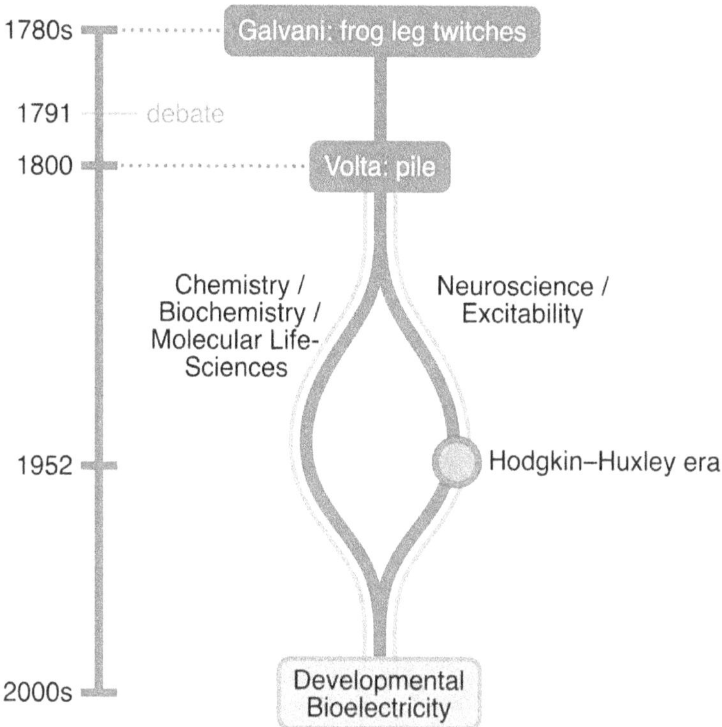

Once the split is named, it becomes easier to notice how it shaped intuitions. A neuroscientist can comfortably speak of voltage as causal: change the voltage across a membrane and you change whether a neuron fires. A molecular biologist, trained on pathways and transcription factors, often treats voltage as downstream: a consequence of metabolism or ion balance, not an information-bearing variable in its own right. The same membrane that looks like a computational device to one community can look like housekeeping to another.

The deeper point is that Galvani's intuition did not require him to know about sodium channels, potassium gradients, or ATP-driven pumps. He did not need the details to recognize the category of phenomenon: living tissues store electrical energy and can release it in organized ways. When he saw a frog leg kick at the touch of metal, he was seeing an output of a hidden architecture. That architecture is not confined to brains.

A concrete way to feel this is to imagine repeating Galvani's experiment with modern equipment, but keeping the spirit of curiosity intact. Suppose you take an excitable muscle preparation and stimulate it with a brief current pulse. You see contraction. Now you take a patch of ordinary epithelial tissue—skin cells, for instance—and you measure its electrical state without stimulating it at all. You find that it sits at a stable membrane voltage. It is not spiking dramatically, but it is not electrically inert. It is maintaining a separation of charge across a thin membrane, like a charged capacitor. It is doing so actively, spending energy to keep ions distributed unevenly. The electrical state is part of what it means for the tissue to be alive. Galvani's frog leg made the electrical nature of life *visible* by producing motion; but visibility is not the same as importance.

This returns us to the debate itself, and to a gentler reading of it. Galvani emphasized the animal; Volta emphasized the metals. Each was tracking a real mechanism. Volta's emphasis helped build the physical sciences of electricity and the technologies that followed. Galvani's emphasis planted a seed that would take much longer to sprout: the notion that physiology includes an electrical dimension that can be primary rather than merely responsive.

The seed did sprout, though often in unexpected places. The language of "resting potential" and "membrane potential" turned animal electricity into something quantitative. The idea of ion channels turned it into something mechanistic. More recently, the observation that patterns of voltage across tissues can correlate with growth, regeneration, and tumor-like behavior has revived the older, broader implication: that electricity is not only a carrier of fast messages in nerves but also a medium for slower instructions in development.

None of that diminishes genetics. Genes build the proteins that form ion channels and pumps; genes determine cell types that interpret signals differently; genes constrain what patterns are possible. But the historical split encouraged a false choice: either life is written in chemicals, or life is animated by electricity. A frog leg, suspended between two metals, suggested something more interesting. Chemistry can create electrical gradients; electrical gradients can guide chemistry. The living body is where those two descriptions cease to be rivals and become two faces of the same organization.

Galvani never had the satisfaction of writing down a membrane equation or measuring a millivolt in a single cell. What he had was the stubborn evidence of movement: a piece of life responding to invisible forces in a way that looked less like passive matter and more like a prepared system waiting for a signal. The frog's kick was not a metaphor; it was a measurement. It said that, under the skin of biology, there is an electrical order that can be disturbed, redirected, and perhaps, with enough understanding, persuaded.

A frog leg on a bench in Bologna could not settle the argument of what electricity *is*. It did something more enduring: it made it impossible to pretend that the question does not belong to biology.

1.2 The Silent Humming of the Body

The nervous system has a public relations problem, and it is winning. Ask a roomful of people what "bioelectricity" means and most will picture a lightning-fast nerve impulse, a jagged spike on a monitor, perhaps a heart trace in a hospital drama. Those images are not wrong; they are simply parochial. They are the fireworks of physiology: bright, brief, and easy to photograph. Beneath them lies a quieter electrical life that rarely earns an oscilloscope screenshot, yet occupies far more of the body's real estate.

A useful starting point is a small act of imagination. Picture a neuron at rest, holding itself around −70 millivolts relative to its surroundings. Now picture a skin cell, a liver cell, a bone-forming osteoblast, or a cell lining your gut. The temptation is to imagine them as electrically blank, like bricks in a wall: chemically busy, perhaps, but not part of the electrical story. In reality, those cells also sit at a voltage. They, too, maintain a charge separation across their membranes. They do not advertise it with frequent spikes, but their electrical state is no less real for being steady. It is more like a low continuous note than a drumbeat: a *hum*.

The technical term for that charge separation is the *membrane potential*, often written as V_{mem}. It is the voltage difference between the inside of a cell and the outside. A voltage, in everyday terms, is a measure of how much electrical "push" exists to move charge from one place to another. Cells create this push by keeping different concentrations of ions—charged atoms like sodium (Na^+), potassium (K^+), chloride (Cl^-), and calcium (Ca^{2+})—on either side of their membranes.

In neurons, changes in V_{mem} can be rapid and dramatic. In most other cells, V_{mem} tends to shift more slowly, and often by smaller amounts, but those shifts can still have consequences. They can change how cells divide, how they move, how tightly they cling to neighbors, what genes they express, and how they interpret chemical signals. Electricity here is not a messenger that carries a sentence from one end of the body to the other in milliseconds; it is part of the cell's *setting*, like temperature or pH, a background condition that biases what the cell is willing to do.

That background is not free. A cell does not merely inherit its voltage the way a rock inherits its color. It must work to hold the line. The main piece of machinery responsible is a protein embedded in the membrane called the sodium–potassium pump, or Na^+/K^+-ATPase. It spends chemical energy (from ATP, the cell's energy currency) to push ions against their natural tendencies. In each cycle, it typically moves three Na^+ ions out of the cell and two K^+ ions into the cell. That arithmetic matters. Sending more positive charge out than comes back in leaves the inside of the cell slightly more negative, making the pump *electrogenic*: it contributes directly to voltage.

The phrase "pump" can mislead, because it suggests a mechanical piston. A better image is a vigilant doorman with a strict counting rule. The doorman has to be paid, and the body pays constantly. In the brain, estimates from classic neurobiology texts put the energetic cost of maintaining these ion gradients at a startling fraction of the total budget, on the order of tens of percent. Even when you are doing nothing in particular, a great deal of metabolic effort goes into preventing your cells from drifting into electrical sameness.

Oddly, in many cells the Na^+/K^+ pump's *direct* electrical contribution to the resting voltage is modest—sometimes described as about a millivolt in well-studied nerve preparations—and yet its *indirect* contribution is enormous. By maintaining the sodium and potassium gradients, it sets up the stored energy that other channels can tap. That is the quiet logic of the hum: the pump maintains the reservoirs; the channels open and close to let charge flow; the voltage is the bookkeeping that tells you how far from equilibrium the cell is being held.

This is the point where the neuron-centric story becomes misleading. The electrical personality of a neuron is dominated by specialized voltage-gated channels that open explosively, producing spikes. But most cells are not trying to transmit fast messages across centimeters of tissue. A skin cell's job is to form a barrier, coordinate with neighbors, respond to damage, and help rebuild. A liver cell's job is to run a chemical factory, detoxifying compounds and managing energy storage. Bone cells must sense mechanical load and orchestrate mineral deposition and remodeling. None of these requires millisecond spikes as a primary behavior. What they require is stable coordination across many cells over minutes, hours, and days. That kind of coordination is compatible with electrical signals that look, to an electrophysiologist, almost boring: gentle shifts in V_{mem}, persistent voltage differences between regions of tissue, and slow waves that spread not by rapid firing but by shared ionic states.

A historical anecdote makes this less abstract. Long before developmental biologists had fluorescent reporters or genetic knockouts, physiologists noticed something peculiar about wounds. If you cut skin, the cut does not simply sit there waiting for chemical factors to diffuse in. Cells at the edge migrate with direction and purpose, the tissue closes, and the whole area reorganizes. In the nineteenth and early twentieth centuries, experimenters began measuring electrical currents associated with injury. The details and instruments evolved, but the recurring observation was that intact epithelia (the sheets of cells that line skin and organs) maintain a kind of transepithelial voltage: one side of the sheet sits at a different electrical potential than the other because ions are pumped and leaked in a coordinated way. When you break that sheet, you create a local short circuit. Current flows, and an endogenous electric field forms around the wound.

That field is not the cartoon electricity of sparks; it is gentle, physiological, and persistent. Yet it can bias cell behavior. In many wound-healing models, cells migrate directionally in response to fields of a strength comparable to what wounded tissue naturally produces. The phenomenon is called *galvanotaxis* or *electrotaxis*: movement guided by an electric field. It has an uncanny implication. A wound is not only a biochemical crisis; it is also an electrical event. The body notices damage partly because damage distorts its electrical geometry.

Once you accept that, the skin ceases to be an inert wrapper. It becomes an electrical organ. Not in the sense that it produces shocks, but in the sense that it maintains voltage patterns across a living sheet, and those patterns can be perturbed in ways that matter.

The liver, too, has an electrical life that is easy to overlook because it does not contract or think. Hepatocytes (liver cells) are polarized: they have distinct "sides" that face different environments, such as bile canaliculi on one side and blood on another. Maintaining that polarity involves controlled ion transport and membrane potentials. The liver's job includes moving charged molecules and salts in and out of compartments; electrical gradients are part of how those movements are organized. When liver tissue is injured, it must coordinate repair, proliferation, and remodeling. Chemical signals are essential, but they act in a context where every cell has an electrical state that can change with metabolism, ion transporter expression, inflammation, and tissue architecture.

Bone seems even less electrical at first glance. Bone is hard, mineralized, and slow. Yet bone is filled with living cells embedded in a matrix, constantly sensing strain and repairing microdamage. Osteoblasts build bone; osteoclasts resorb it; osteocytes act as mechanosensors. Electrical phenomena have long hovered around bone biology, partly because bone exhibits piezoelectric-like behaviors: mechanical stress can generate small electrical potentials in collagen-rich structures. Even leaving that aside, bone cells are still cells. They maintain membrane potentials, express ion channels, and respond to ionic environments. Calcium ions, in particular, are not only building blocks for hydroxyapatite; they are also electrical signals, and the movement of calcium across membranes is inseparable from voltage.

A common reaction to all of this is to ask whether the voltage is a cause or a consequence. Does a tissue heal because of its electric field, or does the field arise because the tissue is healing? In living systems, those directions often braid together. Ion transport creates voltage; voltage changes ion transport; both influence signaling pathways. The trick is to recognize that calling electricity a mere byproduct is often a way of avoiding the harder question: whether the electrical state carries *information*.

Information is a slippery word outside communication theory. Here it can be taken in a modest sense: a pattern that can be read and acted upon. A voltage difference across a tissue is a pattern. If cells change behavior in response to it, then it is information in action, even if it is not encoded as a discrete message.

A good dinner-party test is to ask what would happen if you could nudge that pattern. In the nervous system, nudging voltage is obviously causal; it can trigger a spike or silence it. In non-neural tissues, nudging voltage is less theatrical but can be just as consequential. Experiments from Michael Levin's lab and others over the past couple of decades have shown that changing bioelectric states in embryos and regenerating animals can alter patterning outcomes in ways that are difficult to explain as mere side effects. The details vary by organism and method, but the recurring theme is that altering ion channel activity, gap-junction connectivity (the tiny conduits that let ions and small molecules pass directly between neighboring cells), or the resting membrane potential of groups of cells can shift developmental decisions. Voltage becomes a kind of collective variable: not a property of one cell in isolation, but a state that can spread across a tissue.

To avoid making this sound like an alternative to genetics, it helps to think about how different kinds of instructions coexist in your body. Genes specify the parts list and many of the rules. Chemical gradients and signaling molecules provide local cues. Electrical states provide a fast, physically grounded way to couple distant regions and synchronize behavior without requiring a dedicated nerve. A sheet of epithelium connected by gap junctions can behave like an electrical fabric: perturb one region and the change can propagate as ions redistribute, without any cell needing to "know" the whole plan in molecular detail.

This fabric also has memory. Not memory in the psychological sense, but in the dynamical sense: systems can have stable states. A pattern of membrane potentials across a tissue can settle into a configuration and resist perturbation, like a ball in a valley. That stability can be biologically valuable. It can allow a tissue to maintain polarity, boundaries, or a readiness to respond. When stable states are disrupted, the consequences can include inappropriate growth or failed repair.

Cancer offers a sobering example of why the hum matters. Tumors are often discussed as genetic diseases, and in many respects they are: mutations can unleash proliferation. Yet cells do not proliferate in a vacuum. They respond to their electrical and mechanical environment as well as to growth factors. Many cancer cells show altered membrane potentials compared to their healthy counterparts; they can be relatively depolarized (less negative inside), and that electrical shift correlates with changes in cell cycle behavior and differentiation status in multiple models. The point is not that voltage replaces mutations as an explanation, but that the electrical state can be part of the control system that decides whether a cell behaves like a well-mannered citizen of a tissue or like a solitary opportunist.

At this stage, a reasonable person might worry that "bioelectricity" is being stretched so far that it means everything and therefore nothing. The antidote is to keep returning to concrete mechanisms and measurable quantities. V_{mem} is measurable. Ion channels and pumps are identifiable proteins with known structures. Electric fields in tissues can be measured with electrodes. Cell migration in response to imposed fields can be quantified. This is not metaphorical electricity; it is physics applied to salty water and lipid membranes.

What makes the body-wide hum so easy to ignore is not its vagueness but its familiarity. We take for granted that cells have inside and outside, that membranes keep order, that tissues coordinate. Electrical gradients are one of the most universal ways living systems impose that order. Life is an improbable island of separation: different concentrations, different compartments, different fates for neighboring cells. Voltage is a signature of that separation.

Neurons dramatize the signature by turning it into pulses. Most of the body plays it as a sustained chord.

Consider, finally, what it means that your skin maintains a transepithelial potential, your liver cells maintain polarity-dependent voltages, your bone cells maintain electrical responsiveness to their environment, and every one of these tissues spends energy to remain electrically out of equilibrium. The body is not a bag of chemicals with a nervous system laid over it like wiring on the surface. It is wiring all the way down: a living structure continuously paying for the privilege of being electrically patterned, continuously generating the quiet conditions that make organized healing, growth, and form possible.

A frog leg twitch can capture attention in an instant. The hum that keeps your tissues poised between stability and change is less arresting, but it is far closer to the everyday miracle: that billions of cells, each a tiny battery, can hold their charge long enough to build a body and keep it whole.

1.3 Miniature Batteries: The Physics of Life

A battery is not defined by its size or its casing. It is defined by an imbalance that has been made stable: a separation of charge and chemistry that can be tapped to do work. In that sense, every living cell is a battery, not as a metaphor of convenience but as a literal physical description. Your cells maintain an electrical pressure difference across a thin membrane, and they spend energy to keep it from collapsing. The voltage is small on human scales—tens of millivolts—but the membrane is thinner than almost any everyday object you can name, only a few nanometres thick. Divide a modest voltage by an astonishingly small distance and you get a strong electric field. Cells live inside those fields.

The non-neural tissues described earlier make this easy to forget because their electrical changes can be slow and subtle. Yet the machinery that makes a neuron excitable is not a special organ bolted onto biology; it is an elaboration of machinery that most cells already have. To understand the hum and the occasional shout, it helps to understand how a cell builds its "charge separation" in the first place.

The starting point is the membrane. A cell membrane is a lipid bilayer: two layers of oily molecules whose water-repelling tails face inward, creating a barrier to most charged particles. Ions, by definition, carry charge and are surrounded by a shell of water molecules in solution. Asking an ion to cross the membrane unaided is like asking someone in a winter coat to squeeze through a wall of oil: energetically expensive and therefore rare. This makes the membrane a good insulator.

Yet a membrane is not an infinite insulator. It is thin, and charges can accumulate on either side. This is exactly what a capacitor does in electronics: two conductors separated by an insulator store separated charge. The cell membrane behaves like a capacitor, and biophysicists often use a standard modelling value of about $1\,\mu\text{F/cm}^2$ for membrane capacitance. That number is not required to appreciate the idea, but it hints at why voltage can change quickly when channels open: a capacitor can be charged and discharged by currents, and the membrane is a capacitor wrapped around the cell.

If the membrane is a capacitor, what provides the current to charge it? Two classes of proteins dominate the answer: *pumps* and *channels*.

Pumps are active machines. They use energy (often from ATP) to move ions against their electrochemical gradients. Channels are passive gates. They open a path through the membrane and let ions flow down their gradients. In everyday language, pumps create imbalance; channels exploit it. The cell's voltage, V_{mem}, is the live financial statement of this arrangement: how much electrical "credit" exists at any moment because the cell has invested energy to keep ions separated.

The most famous pump is the sodium–potassium pump, the Na^+/K^+-ATPase. It does something that sounds simple and is, in molecular terms, astonishingly choreographed. It grabs sodium ions from inside the cell, uses ATP to change its shape, releases them outside, then grabs potassium ions from outside, changes shape back, and releases them inside. The bookkeeping rule is usually three Na^+ out for two K^+ in. This makes the inside slightly more negative and, more importantly, establishes the sodium and potassium gradients that underwrite most of animal electrophysiology.

A common misunderstanding is to imagine the pump as the main source of the resting voltage. It does contribute directly because it is electrogenic, but in many classic nerve preparations its direct effect is small compared to what the gradients enable. The heavy lifting comes from selective permeability. The cell membrane is rarely equally permeable to all ions. At rest, many animal cells are much more permeable to potassium than to sodium because they have "leak" channels that allow K^+ to drift out. When positive potassium ions leave, they leave behind negatively charged molecules that cannot follow easily. The inside becomes more negative. That negativity then pulls potassium back in, and the system settles into an equilibrium where chemical and electrical forces balance. The result is a stable V_{mem}.

This balancing act is why membrane voltage is often described as an *electrochemical* phenomenon rather than merely electrical. The "chemical" part is concentration differences; the "electrical" part is charge. They are inseparable. An ion does not care whether it is being pushed by a concentration gradient or an electric field; it just experiences forces and moves accordingly.

To put a number on this without drowning in equations, it helps to introduce one tool of thought: the *reversal potential*. Every ion species has a voltage at which it would be perfectly content, given its concentration difference across the membrane: a point where the electrical pull exactly cancels the chemical push, so there is no net movement. That voltage can be computed from the Nernst equation, and it turns a concentration gradient into something that looks like a little battery. In circuit models of cells, each ion's gradient is represented as a battery with its own preferred voltage. Channels are like resistors that connect those batteries to the membrane capacitor. Open a potassium channel and the membrane voltage gets tugged toward potassium's preferred value. Open a sodium channel and the voltage gets tugged toward sodium's preferred value.

The resting membrane potential is, in this picture, a compromise. It sits near the reversal potential of the ion to which the membrane is most permeable at rest, often potassium, but it is shifted by other permeabilities. David Goldman provided a key theoretical step in 1943 by developing a constant-field approach that leads to what is now known, in its later form, as the Goldman–Hodgkin–Katz (GHK) equation: a way to estimate membrane potential from the relative permeabilities of multiple ions. The GHK framework is powerful because it makes clear that V_{mem} is not a mystical property of "life force" but a predictable outcome of ion concentrations and membrane permeability.

The mid-twentieth century transformed this from theory into a mechanistic story of excitability. Hodgkin and Katz (1949) demonstrated that the action potential in the squid giant axon depends critically on sodium in the surrounding solution, helping to cement the ionic view of bioelectricity. Hodgkin and Huxley (1952) then produced a quantitative model in which the membrane behaves like a capacitor in parallel with variable conductances for sodium and potassium. Their work did not merely describe spikes; it made spikes inevitable given certain kinds of channels. If you build a membrane with voltage-dependent gates that change permeability, a spike is not magic. It is physics with feedback.

These historical moments matter beyond neuroscience because they give a template for thinking about any cell. Once a membrane is treated as a capacitor and channels as conductances, the quiet hum of a skin cell and the sharp shout of a neuron differ mostly in parameter settings: which channels are expressed, how strongly they respond to voltage, how cells are coupled to neighbors, and what gradients the pumps maintain.

A concrete example brings the machinery into view. Consider what happens when a neuron fires an action potential. At rest, the neuron is more permeable to potassium, so V_{mem} sits closer to potassium's reversal potential. Then a stimulus opens voltage-gated sodium channels. Sodium floods in because sodium is high outside and low inside, and because the inside is negative. This influx depolarizes the membrane, which opens even more sodium channels. It is a positive feedback loop: voltage opens channels, channels change voltage. The spike rises rapidly. Then sodium channels inactivate and potassium channels open, letting potassium flow out, repolarizing and often hyperpolarizing the cell. The membrane capacitor charges and discharges through these conductances. Throughout this whole drama, the Na^+/K^+ pump is not the sprinter; it is the long-distance support crew, restoring gradients over time so the neuron can fire again.

Now take away the voltage-gated sodium channels, or make them rare, and you largely remove the fireworks. The cell still has a voltage. It may shift slowly as pumps and leak channels adjust, as metabolic state changes, as neighboring cells tug on it through gap junctions. Those slow shifts can be biologically meaningful even without spikes, because a voltage is a control knob that influences many proteins, including voltage-sensitive channels, transporters, and sometimes even transcriptional pathways indirectly via calcium and other messengers.

The battery analogy becomes most vivid when one remembers what a pump is doing energetically. Moving ions against gradients stores free energy in the distribution of those ions. It is as if the cell is lifting weights and placing them on shelves at different heights. The shelves are the inside and outside of the membrane; the heights are chemical potentials. When a channel opens, the weights can fall, and the falling can power work: moving other molecules, changing cell volume, triggering signalling cascades, altering the cytoskeleton, or simply changing voltage in a way that the cell interprets.

A schematic can help keep the cast of characters straight.

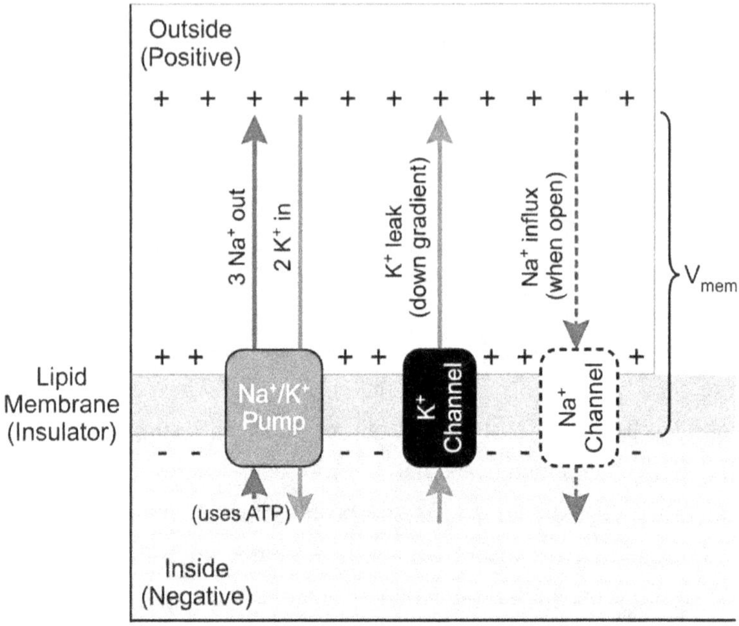

Mechanism: The Na^+/K^+ pump actively creates ion gradients. K^+ leak channels allow potassium to flow out, establishing the resting membrane potential. Na^+ channels allow influx when gated open.

The diagram is necessarily simplified. Real membranes host dozens of pumps, exchangers, and channels, and ions rarely act alone. Chloride often follows, sometimes passively, sometimes via transporters that actively set its distribution. Calcium is kept extremely low inside most cells, which makes calcium influx a potent signal. Proton pumps and bicarbonate transporters help regulate pH. But the simplification is honest in its essence: a cell is an energy-consuming device that maintains gradients, and voltage is the immediate consequence.

A second simplification hides in the very phrase "inside negative, outside positive." The bulk fluids inside and outside a cell are nearly electrically neutral. The charge separation that creates V_{mem} exists in thin layers right at the membrane surfaces, like the charged plates of a capacitor. That is why a small number of ions can create a measurable voltage without making the whole cell a charged object that would repel its neighbors like static-cling balloons.

The physics also explains why bioelectricity is such a tempting candidate for coordination in development and regeneration. Chemical gradients can diffuse and be degraded; they can be buffered and bound. Voltage patterns can also change, but they have a distinctive kind of immediacy because electric forces act on all charges, not just on one ligand and its receptor. An electric field can bias the movement of ions and charged proteins, can influence the opening probability of voltage-sensitive channels, and can alter the distribution of signalling molecules that carry charge. This does not mean electricity is always the dominant cue, only that it is a cue with a different set of strengths than chemistry alone.

There is a further, subtler point: the cell's electrical state is not a mere readout of what the cell is doing; it is part of what the cell *is*. A cell type can be characterized not only by which genes it expresses but by its typical membrane potential and its typical repertoire of channels and pumps. Differentiation often involves changes in ion channel expression. Proliferating cells and differentiated cells can differ in their electrical profiles. In that sense, electrical state belongs in the same family of descriptors as metabolism, shape, and gene expression.

This framing also rescues Galvani from the caricature of having been "wrong." His animal electricity was not a special fluid. It was, as later work made clear, the electrical behavior of ions in and around membranes. Volta was not wrong either; his metals did generate electricity, and his pile showed how. The deeper resolution is that life sits at the intersection of electrochemistry and organization. Cells are not passive sacks of solution waiting for a shock; they are dynamic batteries built from proteins, lipids, and salt water, tuned by evolution to store and release energy in specific ways.

A final concrete image brings the argument back down to the scale of touch. Hold a AA battery in one hand. It sits quietly, doing nothing visible. Its power is potential, not performance. Your cells are like that. Most of the time, they are not firing or contracting, but they are *ready*. Their membranes hold charge. Their pumps keep gradients steep. Their channels stand guard, some leaky, some gated, some exquisitely sensitive to voltage, stretch, chemicals, or temperature. When conditions change—a wound opens, an embryo reorganizes, a neuron receives synaptic input—the stored energy becomes action.

The astonishing part is not that a neuron can spike, but that the body can afford to keep so many microscopic batteries charged, all the time, across a lifetime, in tissues that will never appear on an EEG screen. The hum is not incidental. It is the price of staying alive in a world where equilibrium is death.

Chapter 2

The Invisible Architect

Long before a physical organ takes shape, an electrical scaffold outlines its future geometry in the dark silence of the embryo. We delve into the hidden conversations occurring between cells, discovering how voltage gradients act as the master builders directing the construction of complex living forms.

2.1 Hardwiring the Collective

A single cell is a complete life. It eats, senses, defends itself, divides, and dies. Even in the body, most cells retain something of that ancient independence: each has its own boundary, its own chemistry, its own metabolism, its own electrical state. Yet an animal is not a pile of cells. It behaves as a coordinated agent, building a face in the right place, repairing a cut without making a second mouth, keeping a heart beating in synchrony, and stopping growth at a sensible size. The puzzle is not that cells can do impressive things. The puzzle is how millions of impressive little lives agree on one shared plan.

One answer comes from a piece of cellular hardware so small and so ordinary-looking that it was easy to treat as mere plumbing: the gap junction. Under a microscope, a gap junction is not a nerve, not a hormone gland, not a signaling molecule drifting through space. It is a literal bridge. Two neighboring cells press their membranes close together and dock matching protein channels, forming a pore that spans both membranes. Ions and small molecules can then pass directly from one cytoplasm to the next, without asking permission from the extracellular world. Where a cell membrane usually enforces privacy, the gap junction creates a kind of cytoplasmic commons.

If the phrase "cell-to-cell channel" sounds mundane, consider what it accomplishes. A membrane is an electrical device. It separates charge, it hosts pumps and channels, it supports a voltage difference (the resting membrane potential, often abbreviated V_{mem}). When two cells are electrically isolated, each cell's voltage is largely its own affair, shaped by its own ion channels and pumps, its own metabolic state, its own local signals. When cells are joined by many gap junctions, they no longer inhabit separate electrical worlds. The voltages can influence each other. Local disturbances can propagate. A pattern can extend across a tissue. What looks like a sheet of independent units begins to act like a network.

A useful way to picture this is to imagine each cell as a house with its own battery and its own wiring. Without connections, each house can turn lights on and off, charge devices, and run appliances, but the neighborhood has no shared grid. With gap junctions, the houses connect their circuits. Now, the neighborhood can sustain a voltage landscape: not one cell's voltage, but a collective distribution of voltages. The crucial shift is conceptual. The "state" that matters is no longer just intracellular; it becomes *intercellular*. Information is no longer confined to a cell; it can be stored and communicated in the pattern of electrical coupling itself.

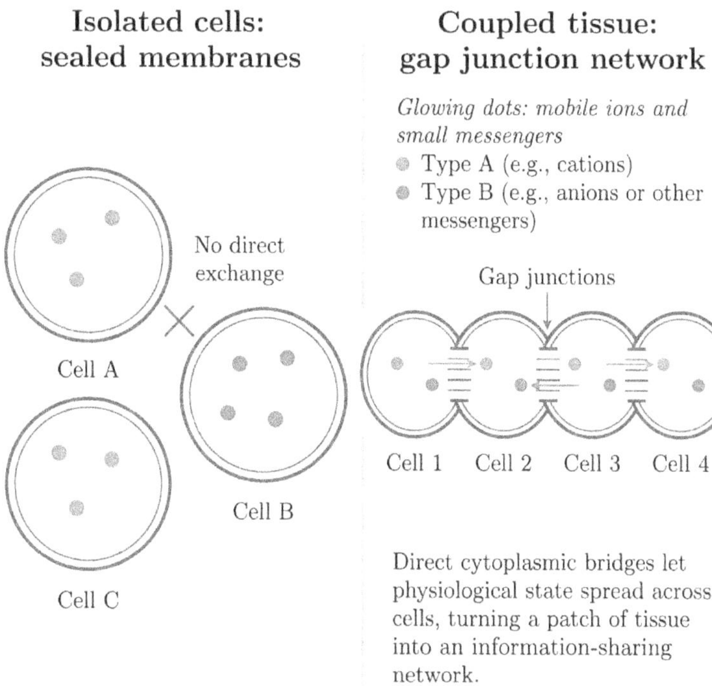

The first temptation, especially if one has been trained to think of biology as chemical, is to treat gap junctions as a convenience for passing nutrients or clearing waste. They do that, but their deeper role is that they couple *physiology*. Not just matter, but state. A cell's resting potential is shaped by the balance of ions across its membrane. When ions can move between cells, a cell's resting potential becomes partly a function of its neighbors. This is what it means to build a collective: the boundaries remain, but the private variables leak into a shared space.

One can see why multicellular life would seize on this. A body has problems that no single cell can solve. Consider scale. A cell can decide whether to divide, migrate, or differentiate, but it cannot, by itself, know whether the organism is the right size, whether a limb is proportionate, or whether a wound has been closed with the correct contour. Those are global questions. Chemical morphogens diffusing through tissue can certainly broadcast positional information, but diffusion is slow and local, and chemical concentration fields are vulnerable to noise and to the turbulence of living tissue. Electrical coupling offers a complementary channel: a way to coordinate without moving proteins long distances, and a way to integrate signals across a field of cells.

The story becomes more concrete if we linger on what can pass through a gap junction. These pores are selective by size and charge; they do not freely exchange large proteins or DNA. What they do exchange readily includes ions like K^+, Na^+, Cl^-, and small signaling molecules such as cyclic nucleotides, inositol trisphosphate, and sometimes even small metabolites. That list matters because these are precisely the ingredients that set electrical state and trigger downstream biochemical cascades. A pulse of Ca^{2+}, for example, is not merely a chemical event; it is a currency of cellular decision-making, tightly coupled to membrane voltage. If gap junctions allow one cell's physiological shift to nudge another's, then the tissue can behave less like a crowd and more like a choir.

There is also an underappreciated social dimension to this plumbing. Multicellular life requires a kind of enforced cooperation. In a unicellular world, a cell that grabs resources and divides aggressively is, from its own point of view, doing brilliantly. In a body, that same strategy is cancer. The body's success depends on each cell restraining certain selfish impulses and synchronizing its behavior with a plan that exists at the scale of tissue and organ. Electrical coupling is one of the ways cells are made answerable to their neighbors. It is harder to be a lone wolf when your internal conditions are literally connected to the group's.

The molecular history behind gap junctions provides a small but telling anecdote about how science learns what matters. When early microscopists began to see cell boundaries, the membrane became a symbol of individuality: a line that defined self versus outside. Much later, when electron microscopy and careful physiology revealed tiny "junctions" where membranes came into close apposition, the field had to adjust. The cell was not always an island. In the 1960s and 1970s, a wave of work in many tissues showed that cells could be electrically coupled, and that current could spread from one to another without synapses in the neuronal sense. The discovery did not dethrone chemistry; it complicated it. Cells were not just bags of reactions in parallel. They were nodes in circuits.

The human heart offers an intuitive example, even if it is not the most developmentally exotic. Heart muscle cells contract in a coordinated wave because they are linked by gap junctions at specialized contact points (within structures called intercalated discs). The "spark" of depolarization spreads from cell to cell; the tissue behaves like one functional unit. When this coupling is disrupted, rhythm disorders can emerge. The heart makes visible a general principle: coupling turns a sheet of individual excitable units into something that has a collective behavior.

Developmental tissues are often not "excitable" in the dramatic, spiking way nerves and muscles are, but they still carry voltages and currents. Their electrical language is slower and more analog: gradients of resting potential rather than millisecond action potentials. This can make it easy to miss, because it does not announce itself with a twitch or a firing neuron. Yet slow does not mean trivial. A tissue that maintains a stable voltage pattern for hours is doing something that resembles memory: holding a physiological state long enough to guide growth and differentiation.

To make sense of this, it helps to distinguish two kinds of "brains." One is the neural brain, built of neurons with fast electrical spikes and well-mapped synaptic connections. The other is a broader, older system: the electrically coupled somatic tissue that communicates through resting potentials and gap junctions. The second is not a metaphorical brain in the sense of conscious thought, but it is a real information-processing network. It collects inputs, integrates across space, and produces coordinated outputs. In a developing embryo, it is often operating before the nervous system has even formed.

What does "information processing" mean at this level? Not abstract symbols, but reliable mappings from state to action. A cell that is more depolarized or more hyperpolarized (more positive or more negative inside relative to outside) can change what genes it expresses, how it moves, whether it divides, and what signals it emits. Those choices are not arbitrary; they follow from voltage-sensitive proteins, voltage-gated channels, electrogenic transporters, and second-messenger cascades. If one cell's voltage can influence a neighbor's voltage via gap junctions, then the tissue can coordinate choices across many cells. The result can be a boundary between domains (a future organ field), a directional bias (polarity), or a synchronized shift (collective differentiation).

A concrete developmental example comes from amphibian embryos, where Michael Levin's lab and others have used voltage-sensitive dyes and genetic or pharmacological manipulation to show that resting potential patterns appear early and matter. In *Xenopus laevis*, work by Pai and colleagues (2012) showed that a distinct cluster of hyperpolarized cells appears in the region that will give rise to the eye, before the eye is visible. When those cells are experimentally depolarized, eye development is disrupted in characteristic ways. Even more strikingly, when cells in other locations are forced into an eye-like electrical state, ectopic eye structures can sometimes be induced in unusual places. The point is not that voltage replaces genes; the point is that physiology can act as an instructive layer that helps decide what gene programs get stabilized where.

Gap junctions are essential to making such patterns into tissue-scale decisions rather than isolated curiosities. Without coupling, one cell's electrical state can be a private quirk. With coupling, a cluster can become a domain. The tissue can act like a fabric: tug one corner, and the tension redistributes. In physiology, the "tension" is ionic and electrical, and it is constantly being maintained by pumps and channels, which means it can be both robust and adjustable.

There is a further, slightly unsettling implication: if tissues are electrically coupled networks, then a developmental defect is not always a problem of a single cell doing the wrong thing. It can be a network-level failure, like a short circuit, a broken connection, or a corrupted pattern. In that light, some congenital anomalies and perhaps even some cancers begin to look less like purely genetic misprints and more like disorders of collective state. Levin and others have argued, on experimental grounds, that depolarization states can correlate with tumor-like behavior in some models, and that restoring normal bioelectric patterns can nudge cells back toward orderly behavior. Whether this becomes a major therapeutic paradigm remains to be seen, but the conceptual shift is already valuable: "control" in biology is not only molecular; it is also electrical and networked.

To keep this grounded, it is worth asking what prevents a coupled tissue from dissolving into mushy sameness. If ions and small messengers spread through gap junctions, why don't all cells become identical? The answer is that coupling is rarely uniform. Gap junctions open and close; different connexin proteins (or innexins in many invertebrates) create channels with different properties; tissues can be partitioned into coupling compartments; and, crucially, each cell still has its own membrane machinery that can maintain or resist changes. Coupling is a negotiation, not a merger. It allows the tissue to share a physiological context while still permitting local specialization.

This balance between shared context and local identity is one of the deepest themes of multicellular life. It is what allows an epithelium to behave as a cohesive sheet while still letting one region become an eye and another become a mouth. It is what allows wounds to recruit a coordinated response without turning the whole body into scar tissue. It is what allows growth to stop when the shape is correct. A purely isolated assembly of cells would struggle to sustain coherence; a perfectly fused syncytium would struggle to sustain such diversity. Gap junction networks sit in the productive middle.

If this sounds like a political arrangement in miniature, that is not an accident of language. Cooperation always involves a trade: freedom for coordination. In tissues, the trade is conducted in volts and ions rather than laws and norms, but the logic is recognizably similar. A cell that is coupled to its neighbors is less sovereign, but it gains access to a larger intelligence: information about the state of the group, the direction of collective change, and the constraints of a shared plan. The astonishing thing is that this is not merely an abstract metaphor. It is hardware, built into the membranes, opening and closing thousands of times a second, letting tiny charges drift across microscopic bridges so that a body can behave like a single, persistent self.

2.2 The Electric Scaffold

A tissue wired together by gap junctions can behave like a single physiological unit, but wiring alone does not explain form. A connected city still needs zoning. If multicellular life is going to build an eye here and a mouth there, it needs a way to mark regions long before those features exist as bumps, folds, or pigment. This early marking is easy to imagine in chemistry: gradients of morphogens, local gene expression domains, transcription factors turning on and off. It is harder to imagine in electricity, because "electricity" sounds like something that happens fast and locally, a spark between points. Development is slow. An embryo can take hours just to decide where a face will go.

The surprise is that electrical patterns in embryos often *are* slow and structured. They look less like lightning and more like a landscape: stable voltage differences across a sheet of cells. The key quantity is again the resting membrane potential, V_{mem}: the voltage across each cell's membrane, set by ion channels and pumps. Each cell has a V_{mem}, but because tissues are coupled, the meaningful object is a *map* of V_{mem} across the embryo. In that map, some regions are relatively more negative inside (hyperpolarized) and others relatively less negative (depolarized). These differences can form boundaries and fields that foreshadow anatomy.

Calling this an "electric scaffold" is not a poetic flourish. A scaffold is something you build first because it constrains what comes next. Voltage patterns can appear before a structure is visible and before some of the familiar genetic markers are expressed in their final configuration. They constrain what the tissue is likely to do: where it will proliferate, where it will invaginate, where it will differentiate, where it will resist change. In other words, they are not merely a readout of development already underway. They can be part of the mechanism that makes development orderly.

For most of the history of embryology, this claim was difficult to evaluate because the embryo's electric life was hard to see. One could insert microelectrodes into single cells, but that gives point measurements: here, in this one cell, the voltage is such-and-such. A voltage landscape is a picture, not a list. It requires many measurements, coordinated in space, ideally in an intact, living embryo. That desire met an older dream in physiology: to make electricity visible the way a stain makes a tissue visible.

The historical anecdote is instructive because it shows how a scientific field changes when a new kind of seeing becomes possible. In the mid-twentieth century, electrophysiology was dominated by exquisite measurements in excitable cells: squid giant axons, frog neuromuscular junctions, mammalian neurons. The tools were sharp, the signals were dramatic, and the conceptual center of gravity shifted toward fast spikes and synapses. Meanwhile, developmental biologists mapped gene expression and morphogen gradients, and the embryo became a theatre of chemical signaling. Endogenous electrical patterns in non-neural tissue were often relegated to the margins, partly because they were difficult to measure cleanly and partly because they were easy to dismiss as side effects of metabolism.

Then optical reporters arrived. Voltage-sensitive dyes and, later, genetically encoded voltage indicators made it possible to image membrane potential in space and time. These tools are not perfect: they can be noisy, phototoxic, and tricky to calibrate; they report relative changes more reliably than absolute millivolts; they can be influenced by tissue thickness and dye loading. But they offer something microelectrodes cannot: a way to see the embryo's electrical organization as a whole. The embryo stops being a collection of voltage readings and becomes a picture with shapes.

The picture is often startlingly anatomical.

In frog embryos, one can image regions that will later become facial features and see them outlined in voltage before they exist physically. The cells are still a smooth surface. No eye socket, no mouth opening. Yet the voltage map can show sharp contrasts that trace where those structures will form. Michael Levin's lab and others helped popularize the phrase "prepattern" for this phenomenon: a pattern that exists early and predicts a later anatomical pattern. "Predicts" can sound weak, as if the embryo were merely forecasting. The more striking claim is that the prepattern can be *instructive*: perturbing it can change what the embryo builds.

A concrete example comes from *Xenopus laevis* eye development. Pai and colleagues (2012) reported that a cluster of cells in the anterior region becomes distinctly hyperpolarized before the eye primordia are morphologically apparent. When that hyperpolarization is disrupted by experimentally depolarizing those lineages, eye formation is impaired in reproducible ways, including smaller or malformed eyes. Even more provocative, imposing an eye-like V_{mem} state in cells outside the normal eye field can sometimes trigger ectopic eye structures in unusual locations. This is not magic, and it is not a claim that voltage alone writes an entire organ out of nothing. Ectopic organs are constrained by competence, by available gene networks, by tissue context. But the experiments make a powerful point: a physiological state can carry organ-level meaning.

How can a difference of a few tens of millivolts, held across membranes thinner than a soap bubble, possibly influence something as grand as a face? The mechanism is not a single wire from voltage to "make eye." It is a web of transduction. Voltage influences the movement of ions such as Ca^{2+}, and Ca^{2+} is one of the cell's most versatile internal signals, touching gene expression, cytoskeletal dynamics, secretion, and cell cycle. Voltage can regulate transporters that import or export small molecules, including neurotransmitters and other signaling compounds used well outside the nervous system. Voltage can alter the distribution of charged molecules at the membrane, which changes the behavior of receptors. Through gap junctions, local changes can spread, making tissue-level domains.

Electrical state becomes an input into the same gene regulatory circuits that morphogens and transcription factors feed, but it is an input with distinctive properties: it can be spatially patterned, quickly modulated, and coupled across cells.

It is also an input that naturally lends itself to *boundaries*. In development, boundaries matter. An eye field must be separated from the surrounding tissue, not only chemically but operationally: the cells in the eye field must coordinate with each other and behave differently from their neighbors. Voltage patterns can create such operational boundaries, both by shaping signaling and by shaping coupling. Some regions of tissue are heavily gap-junction coupled; others are relatively isolated. A voltage domain can be maintained as a stable state in a coupled patch, while adjacent patches maintain different states. The embryo can become a quilt of physiological compartments.

The language of "fields" is inevitable here, and it carries philosophical baggage. Older biology flirted with notions of morphogenetic fields that sounded mystical: invisible organizing principles that knew the target form. Modern developmental biology grounded much of this in molecules, gradients, and gene networks. Electrical prepatterns bring the field concept back in a form that is both physically ordinary and conceptually rich. A voltage field is not mystical. It is a measurable distribution of potential differences, maintained by proteins we can name. Yet it behaves like an organizing field in the sense that it coordinates many local actions toward a coherent structure.

Seeing the field requires converting voltage into color, and that translation is what makes the evidence persuasive to a general audience. A voltage map looks like a weather map: bands and hotspots rather than discrete dots. Blue might represent relatively more negative interior (hyperpolarization), red relatively less negative (depolarization). Suddenly the embryo wears a physiological mask. Where a gene expression map might show one transcription factor in a blob and another in a stripe, a bioelectric map can show gradients that do not neatly match known morphogen distributions, as if the embryo were running an additional layer of organization in parallel.

The temptation, at this point, is to say: "So the embryo has an electrical blueprint." That phrase is useful as long as it does not smuggle in the wrong metaphor. A blueprint is static and external, a fixed plan consulted by builders. An embryo's electrical patterns are neither static nor external. They are dynamic and self-generated. They can change with developmental time, respond to perturbation, and interact with gene expression in feedback loops. A better metaphor is a scaffold that can be rearranged as the building grows, with constraints shifting as parts become stable. Even that is imperfect, but it captures the interplay: an early pattern shapes later events, and later events can reshape the pattern.

The methods matter because the claim is extraordinary enough that one should ask how it might be an artifact. Voltage-sensitive dyes report fluorescence changes that correlate with membrane potential, but the signal can be confounded by changes in dye concentration, membrane area, pH, and motion of the embryo. Developmental tissues move and thicken; embryos rotate and twitch. Researchers have had to develop careful controls: ratiometric dyes, immobilization techniques, parallel measurements of calcium dynamics, and functional perturbations that alter voltage in known ways. A particularly persuasive strategy is not merely to image but to *edit* the pattern: to use ion channel drugs, genetic misexpression of channels or pumps, or optogenetic tools that change membrane potential with light, and then to observe whether anatomy changes in a specific, repeatable manner.

That combination of mapping and causal intervention is what elevates electrical prepatterns from an intriguing correlation to a mechanistic hypothesis. If you can see a hyperpolarized patch where an organ will form, and if disrupting that patch disrupts the organ, and if recreating that patch elsewhere can coax organ-like structures in a new location, then you have the outlines of a code. Levin (2013) has argued for a "bioelectric code" in this sense: that specific ranges or spatiotemporal patterns of V_{mem}, together with coupling topology, correspond to specific patterning outcomes. The word "code" is bold, and it invites skepticism, because biology is full of signals that mean different things in different contexts. Yet the experiments force the question: how much anatomical information can be stored in physiology, and how is it read?

One reason the idea is compelling is that voltage patterns are well suited to being both local and global. A transcription factor in one cell does not automatically inform distant cells. A diffusible morphogen can, but only if it is produced and transported in a controlled way, and diffusion blurs boundaries unless the system invests in sharpening mechanisms. Electrical coupling can propagate changes without the slow drift of proteins, and a tissue can maintain gradients and sharp transitions by balancing passive spread with active pumping. The embryo can, in principle, maintain a stable "electric face" even while cells divide, rearrange, and migrate.

The face is not the only place where this logic might apply. Work in *Xenopus* has also implicated endogenous V_{mem} gradients in early brain patterning, with disruptions leading to altered expression of early neural markers and malformed regions. The brain, like the face, is a structure that requires careful patterning across a field: not simply more or fewer cells, but the right kinds of cells in the right arrangement. The idea that voltage gradients could bias Notch signaling, proliferation, and differentiation fits well with what is known about how electrical state can gate biochemical pathways.

At a dinner party, a skeptical friend might ask a fair question: if bioelectric patterns are so important, why did classical embryology succeed without them? The answer is that classical embryology did not ignore electricity so much as lack the tools to treat it as a first-class variable. Developmental biology has always dealt with fields in practice: organizers, competence regions, induction. It has always been about pattern control. Bioelectricity offers a physically grounded way to implement field-like behavior. It does not replace morphogens; it can constrain their production and interpretation. It does not replace genes; it influences which gene networks stabilize in which places. In that sense, it resembles mechanical forces in development: once considered secondary, now known to be deeply entwined with gene expression and tissue patterning.

There is also a human relevance that pulls the topic out of the aquarium of frog embryos. Regenerative medicine and tissue engineering struggle with the same problem the embryo solves effortlessly: not just growing cells, but shaping tissues into functional anatomy. If bioelectric prepatterns help establish organ fields, then learning to read and write those patterns could become a new kind of control knob. Drugs that modulate ion channels are already a mainstay of medicine; they treat arrhythmias, epilepsy, pain, and hypertension. The provocative thought is that some channel-targeting interventions might one day be used not only to alter excitability, but to guide growth and repair, nudging tissues toward correct pattern rather than merely suppressing symptoms.

To keep the provocation honest, it is worth acknowledging how young and technically demanding this science remains. Measuring V_{mem} in vivo across thick, moving tissues is difficult, and linking a voltage pattern to a specific downstream gene program is rarely a single step. The embryo is full of feedback loops: voltage influences gene expression, gene expression changes ion channels, ion channels reshape voltage. Untangling causality requires careful perturbations, and even then the same perturbation can have multiple effects. Yet the presence of these feedback loops is itself a clue that bioelectricity is not just a side effect. Living systems do not build elaborate feedback around irrelevancies.

A remarkable feature of electrical scaffolds is that they are both material and abstract. They are made of ions and proteins, but they have geometry. They define regions, borders, and gradients. They can be visualized, perturbed, and, in some cases, re-written. When a frog embryo carries a faint electrical outline of eyes and mouth before any sculptor's hand has folded the tissue, it suggests that the embryo is not merely constructing parts; it is maintaining a map of where parts ought to be.

The map is not inked on paper. It is sustained moment by moment by pumps spending chemical energy to keep ions out of equilibrium, by channels opening and closing, by cells sharing their state with neighbors. A developing organism is not assembled; it is regulated into existence. The electrical scaffold is one of the ways regulation gains a spatial form: a quiet, persistent pattern that makes the difference between a collection of cells and a coherent body plan.

2.3 Target Codes and Anatomical Memories

A blueprint is only as useful as the builder's willingness to consult it. In living bodies, the more revealing question is not how a pattern is drawn, but how it is *kept*. An embryo is not assembled once and then left alone. Cells divide, migrate, die, and are replaced. Tissues are stressed by motion, infection, and injury. If anatomy were simply the frozen outcome of early events, organisms would be fragile: a small disturbance at the wrong time would derail everything permanently. Yet bodies show a quieter talent that engineers recognize immediately: error correction. They sense deviations from a desired state and act to reduce them.

The familiar example is wound healing. Cut your skin and the gap closes. But "closing" is not the whole story. The skin does not merely seal; it tends, within limits, to restore the right thickness, the right layering, the right barrier properties. Bone remodels in response to load. Blood vessels regrow in patterns that respect local demands. Even at rest, tissues patrol their own integrity, culling cells that misbehave and replacing those that are lost. The organism behaves less like a sculpture and more like a thermostat: a system that maintains a setpoint by continuously comparing what is with what ought to be.

That idea is familiar in physiology (body temperature, blood glucose), but less commonly applied to *shape*. Shape feels like a one-time accomplishment: once the face is built, it is built. Yet development and regeneration keep hinting that tissues store something like a target morphology, a representation of "correct form" that can guide repair and remodeling. The provocative claim of bioelectricity research is that the target is not stored only in DNA and not only in local chemical gradients. It can be stored, at least in part, in the dynamics of an electrically coupled network: stable patterns of resting membrane potential and connectivity that function as an anatomical memory.

The cleanest demonstrations come from animals that regenerate dramatically, because regeneration forces the question in a way ordinary maintenance does not. A salamander regrowing a limb or a planarian worm regrowing a head is not merely patching a hole. It is rebuilding a complex, proportionate structure. Somewhere in that fragment is information about what is missing and what the final structure should be. If you amputate a planarian's head, it regrows a head; if you amputate its tail, it regrows a tail. The fragment "knows" anterior from posterior, and it knows how much to build.

Planarians are especially useful because they are simple enough to manipulate yet complex enough to surprise. They are flatworms with a central nervous system and a remarkable population of adult stem cells. Cut one into pieces and, with enough time, each piece can rebuild a whole worm. For more than a century, biologists have used planarians as a test case for morphogenetic theories. T. H. Morgan, better known for fruit flies and genetics, wrote about regeneration in the early 1900s, grappling with the problem of polarity: why does the correct end form at the correct wound? Even before modern molecular tools, the persistence of polarity across repeated injuries suggested that the animal carries a stable internal coordinate system.

Molecular genetics has identified many chemical gradients and gene networks involved in planarian polarity, but bioelectric work adds an unsettling twist: the target morphology can be *edited* by transiently perturbing electrical communication, and the edit can persist long after the perturbation is gone.

In work from Michael Levin's group and collaborators, including Oviedo and Morokuma (2009), transient disruption of gap junction communication in planarians could induce animals to regenerate with altered body plans, including double-headed worms. The key detail is the word "transient." The worms were not continuously drugged or genetically altered in a permanent way. A temporary blockade of gap junctions during a critical window could flip the outcome of regeneration. More striking still, some of these altered worms retained their new "decision rule" across subsequent amputations: cut a double-headed worm again, and it tends to regenerate as double-headed again. The animal behaves as if its internal map has been rewritten.

This is the kind of result that makes a biologist sit back. If a short-lived physiological perturbation can yield a long-lived change in patterning behavior, then something in the system is functioning as memory. The memory is not merely "scar tissue," because the change persists through extensive remodeling and across regeneration events that rebuild large portions of the body. Nor is it easily dismissed as a one-off developmental accident, because it can be stable and repeatable. One is pushed toward a systems-level picture: the planarian is not executing a fixed script; it is navigating toward a target, and the target itself can be re-specified.

A second line of evidence comes from ion transporters that set membrane voltage. Beane, Morokuma, and colleagues (2011) used pharmacological and genetic approaches implicating an H, K-ATPase in planarian head regeneration. Altering the activity of such pumps changes V_{mem}, shifting the physiological state of tissues. Those interventions can bias regenerative decisions, again suggesting that the "head-versus-tail" choice is not only a matter of local chemistry at the wound but a decision informed by a larger physiological context.

The idea that bioelectric state influences not just *whether* regeneration occurs but *what shape* is restored becomes even clearer when one considers scaling. Regeneration is not simply replacement; it is proportion. A regenerated head must match the size of the body; organs must be scaled appropriately. Beane and colleagues (2013) reported that bioelectric manipulations can alter head and organ size during planarian regeneration without simply reducing overall growth. That distinction matters. If an animal grows less tissue, the head will be smaller as a trivial consequence. But if the amount of tissue produced is similar and yet the final head is proportionally smaller or larger, then the control variable is not growth rate alone. It is an instruction about *target proportion.*

How could an electrical network store a target morphology? A satisfying answer requires a shift in intuition. People tend to think of voltage as a fleeting signal: a spike that passes and vanishes. But tissues can maintain resting potentials as stable states, and networks of cells can maintain patterns of resting potential as stable spatial distributions. In other words, there can be electrical *attractors*: patterns toward which the system tends to return after perturbation.

Modeling work by Law and Levin (2015) on resting potential bistability illustrates a plausible substrate for such memory. The basic idea is that ion channels and pumps can create feedback in a cell's electrical state. For instance, a voltage-dependent channel may open more as the membrane depolarizes, allowing further ion flow that reinforces depolarization; or the opposite, reinforcing hyperpolarization. With the right complement of channels, a cell can have two stable resting potentials, like a switch with two preferred positions. When cells are coupled, the network can support stable domains: regions of one state adjacent to regions of another, with boundaries maintained by the interplay of coupling and local membrane machinery.

This resembles memory in electronics not because biology is secretly a laptop, but because both are physical systems capable of multistability. A flip-flop circuit stores a bit by having two stable states. A tissue could store an anatomical instruction by having stable physiological states distributed in space. The stored information is not "a picture of a head" but a set of constraints: polarity, domain identity, permissible transitions, and the setpoints toward which repair processes converge.

Error correction then becomes a natural consequence. If the tissue has preferred attractor states, disturbances will tend to relax back toward them, provided the system remains within the basin of attraction. That is what homeostasis means in dynamical terms. A wound, for example, is a major disturbance. The tissue's electrical connectivity and voltage patterns are disrupted. But if the network dynamics are organized such that "correct pattern" is a stable attractor, then the wound triggers actions (migration, proliferation, differentiation) that restore both the structure and the underlying physiological pattern.

This way of thinking also clarifies why gap junctions are so central. Gap junction coupling is not merely communication; it is the fabric that allows the system to behave as a coherent dynamical network. It lets local electrical changes influence distant regions, enabling global coordination. If the network is too disconnected, each region may settle into its own local attractor, and global pattern coherence can fail. If it is too connected, distinctions may be washed out. The animal's ability to maintain a body axis and regenerate appropriately may depend on a finely tuned coupling topology: which cells are connected to which, and how strongly, at which times.

At this point, the concept of "anatomical memory" can sound almost intentional, as if the tissue were recalling a past experience. A safer phrasing is that the tissue maintains long-lived internal states that bias future outcomes. Yet the everyday analogy to memory is hard to resist because the behavior is so memory-like. A transient perturbation produces a durable change in what the organism does when challenged again. This is not memory in the psychological sense; it is memory in the control-theory sense: a system variable that retains information over time.

There is another, more human reason to take this seriously. When medicine fails, it often fails at the level of pattern. A wound heals but scars; the tissue closes but does not restore the correct architecture. A nerve regrows but connects incorrectly. A heart remodels after injury in ways that keep it pumping but leave it weaker. Cancer is, among other things, a breakdown of multicellular cooperation: cells cease to respect the organism's pattern-level constraints. If bioelectric networks participate in the enforcement of those constraints, then understanding their "target codes" may open a different approach to therapy. Rather than killing misbehaving cells alone, one might try to restore the pattern-level signals that keep cells behaving as part of a coherent whole.

The phrase "target code" is deliberately ambitious. A code, in the strict sense, is a mapping from symbols to meanings. Here the "symbols" might be voltage ranges, spatial gradients, temporal sequences of V_{mem} changes, or patterns of coupling; the "meaning" would be anatomical outcomes such as "build head here," "stop growth now," "maintain boundary between these tissues." The mapping is unlikely to be simple, and it is certainly context-dependent. But the planarian results make it hard to deny that something like a mapping exists. Different physiological interventions produce different pattern outcomes, sometimes reliably. That regularity is the raw material from which a code could, in principle, be inferred.

There is a practical obstacle to decoding: anatomy is complex, and perturbations are many. The field has increasingly recognized the need for formal representations of morphology and systematic data infrastructure that can connect interventions to outcomes in a way that is searchable and comparable across laboratories. Without such structure, discoveries remain anecdotal, inspiring but hard to generalize. The ambition, ultimately, is not only to marvel at double-headed worms but to build an atlas of pattern control: a way to predict what will happen when a given bioelectric parameter is changed, and to design interventions that steer regeneration toward desired outcomes.

Even without a complete atlas, the control-theoretic framing suggests a more precise way to think about regeneration. A regenerative system requires at least three ingredients: a sensor that detects deviation, an internal representation of the target, and effectors that change tissue state. In biology, these ingredients are distributed. "Sensing" can be mechanical strain, chemical gradients, immune signals, or electrical discontinuities at a wound edge. "Representation" can be gene expression domains, epigenetic states, and physiological patterns such as stable voltage landscapes. “Effectors” are cell behaviors: migration, proliferation, differentiation, apoptosis, extracellular matrix remodeling. Bioelectricity participates in all three, but its special contribution may lie in representation and coordination: a tissue-scale state that can be stable, adjustable, and shared.

Apoptosis, programmed cell death, deserves a place in this story because it often sounds like destruction rather than construction. Yet remodeling frequently requires subtraction. Sculptors remove stone to reveal form. Bodies remove cells to achieve proportion and correct errors. If bioelectric state biases where apoptosis occurs, it can contribute to scaling and shaping without altering the overall capacity to grow. That provides one plausible bridge between voltage patterns and final anatomy: not just “grow here,” but “trim there.”

The philosophical implication is subtle but profound. The gene-centric view of form suggests that morphology is computed locally: each cell reads its genetic program and local signals, and the sum of these local computations yields the global shape. The bioelectric view suggests that there are also global state variables, maintained by networks, that act as higher-level constraints. The organism is not merely a mosaic of local decisions; it is a control system with distributed memory.

To make that less abstract, return to the planarian on the bench. Imagine two worms, genetically identical and living in the same dish. One has the ordinary body plan: one head, one tail. The other has been coaxed, by a brief interruption of gap junctional communication, into becoming double-headed. Now injure them in the same way. They respond differently, not because they have different genes, but because they are in different internal states. The difference is not visible at the wound margin on day one. It emerges as a pattern outcome days later. Something in the worm’s physiology has stored a decision about what "complete" looks like.

That is the essence of anatomical memory: a persistence of target information that survives perturbation and guides future repair. It is also a hint of a deeper unity between development and regeneration. Both require the same miracle, repeated at different timescales: the coordination of countless microscopic acts into a macroscopic form that is not merely produced, but maintained.

A living body is a collective that remembers itself. The memory is not held in a single place, not in a "shape organ," but distributed across tissues as stable patterns of interaction. Electricity, in this context, is not the crackle of nerves but the quiet arithmetic of ions, continuously solving the problem of "what should this be" by comparing the present to an encoded target and pushing, cell by cell, toward the correct geometry.

Chapter 3

Sculpting the Embryo

An embryo is not merely a genetically programmed machine assembling itself by rote; it is an electrically dynamic system where voltage gradients act as the architectural blueprints. This chapter reveals how invisible ionic currents break the symmetry of a fertilized egg and sketch the ghostly outlines of faces and limbs before cells even begin to migrate. We discover that altering these electric scaffolds can rewrite the body plan without changing a single line of genetic code.

3.1 The Spark of Asymmetry

A body that looks roughly symmetric from the outside is quietly lopsided within. The heart sits to the left, the liver mostly to the right; the stomach curves one way, the intestines coil with a consistent handedness. That consistency is so reliable that it can feel like a simple fact of anatomy, as if the embryo merely "follows the plan." Yet for every embryo there is a moment when there is no left or right to follow—no beating heart, no folded gut, not even a recognizable head. There is only a ball of cells dividing with metronomic regularity. Somewhere in that early, apparently featureless time, the organism must decide: *this side will be left.*

Many people learn a standard story of left—right patterning: microscopic whips called cilia create a directional flow of fluid, biasing signals to one side, which then triggers a cascade of genes (often summarized as a Nodal—Pitx2 pathway) that instructs organs to curve, loop, and settle into their places. That story is real and powerful. But it leaves a strangely human question unanswered. How does the embryo know which way is "leftward flow" in the first place? Even before cilia spin, before fluid moves, embryos already carry subtle biases. One of the most unexpected biases is electrical.

Bioelectricity here does not mean nerves firing or muscles contracting. It means the quiet voltage differences that exist across every living cell membrane. A cell membrane is a thin insulating layer separating two salty solutions. Because ions (charged atoms and molecules) are moved unequally across that membrane—mainly by pumps and channels—the inside and outside of a cell sit at different electrical potentials. This difference is the *membrane potential*, often written as V_{mem}. If the inside is more negative than usual, the cell is said to be *hyperpolarized*; if it is less negative (or even positive), it is *depolarized.* In an embryo, cells are not electrically isolated dots. They share ions and small signaling molecules through direct connections called *gap junctions*. The result can be tissue-scale voltage patterns, like a topographic map, spread across an early body that has not yet learned to look like one.

A useful way to picture this is to imagine the embryo as an orchestra tuning up. No melody yet, no recognizable song, only separate instruments testing notes. But the tuning already has structure: certain sections share a reference pitch, certain timbres dominate, and the hall itself shapes what can be heard where. In development, the "tuning" is the set of ion gradients and voltages, and those early biases can set the key in which later genetic and mechanical events will play.

One concrete entry point into this world came from a kind of detective work. In the early 2000s, Dany Adams and colleagues, working with the frog *Xenopus*, conducted broad pharmacological screens: apply a drug that blocks a certain ion transporter, watch what happens to the embryo, and learn which ion flows matter for patterning. This approach can feel blunt, but it has a particular strength: it can reveal crucial steps that happen so early and so subtly that they are hard to spot by anatomy alone.

Among the strongest hits was a proton pump with an intimidating name: the H^+-V-ATPase. It uses ATP (the cell's energy currency) to move protons (H^+ ions) across membranes. Because protons carry charge, moving them creates both pH differences and voltage differences. Adams and collaborators (2006) showed that very early proton flux driven by this pump is necessary for consistent left—right patterning in *Xenopus*. The timing was the shock: the relevant asymmetries appeared during the first few cleavages, at the 4-cell and 8-cell stages—long before the embryo has any visible left or right. Even more surprising, when they disrupted this asymmetry, the embryos did not simply fall apart. Many developed into otherwise healthy tadpoles whose organs were randomly placed: heart on the right in some, normal in others, sometimes more complex rearrangements. That pattern is a signature of an *instructive* signal: the embryo is not merely being poisoned; it is being given scrambled coordinates.

The human face of this problem is a condition called *heterotaxia*, in which internal organs are mirrored or mismatched, sometimes accompanied by congenital heart defects. A complete mirror-image arrangement of the viscera is called *situs inversus*; heterotaxia is often messier, with some organs reversed and others not. The point is not that a frog experiment "explains" every human case. Rather, it reveals an ancient vulnerability: if the embryo's earliest coordinate system is perturbed, later steps can still build organs, but the map that tells those organs where to go becomes unreliable.

How could a proton pump help an embryo decide left from right? The embryo already has another axis: top and bottom (animal and vegetal, in frog terminology), established by the egg long before fertilization. It also has front and back cues that appear early. Left and right, however, are special because they are mirror images: any mechanism that treats them differently must break symmetry in a way that is consistent across embryos. Bioelectric signals are well suited to this because they can do two things at once: create a spatial bias and regulate the movement of charged molecules.

Consider a common developmental signal that is itself a small molecule with charge or that travels in charged forms. A voltage difference can influence where such molecules accumulate, much as static electricity can make hair cling to a comb. A voltage pattern can also gate the opening of certain ion channels, changing calcium influx; calcium, in turn, is a universal messenger that controls enzyme activity and gene expression. Even without invoking exotic "electrical instructions," ordinary cell biology provides multiple routes by which voltage can be translated into biochemistry.

Another strand of evidence made the story more vivid by involving a familiar neuromodulator: serotonin. Many people associate serotonin with mood because antidepressants target serotonin signaling in the adult brain. But serotonin is older than brains. It exists throughout the animal kingdom and plays roles in early development. In work by Takahiro Fukumoto and colleagues (2005), serotonin signaling was implicated as an extremely early step in left—right patterning in both chick and frog embryos. The remarkable claim was not merely that serotonin is present, but that its *localization* and *movement* during early cleavage stages matter for sidedness.

Here gap junctions become central characters. Gap junctions are protein channels that directly connect the interiors of neighboring cells, allowing ions and small molecules to pass from cell to cell. If an early embryo has a network of gap junction connectivity that is not perfectly symmetric—or if a voltage barrier emerges that effectively partitions the embryo into domains—then molecules like serotonin can be shared preferentially on one side. Fukumoto and colleagues connected serotonin's early behavior to both gap junctional communication and ion transporters such as H,K-ATPases (pumps that exchange protons and potassium). In other words, a bioelectric landscape and an intercellular wiring diagram together can decide which cells have access to particular chemical determinants.

The idea sounds almost too abstract until one notices an everyday analogy: cities are not only shaped by where people live, but by where roads allow them to go. In an embryo, the "roads" are gap junctions, and the "traffic rules" can be voltage and pH. A molecule can be made in one place but effectively belong to a region if it can diffuse within that region and is barred from crossing a boundary. Developmental biology has long relied on such compartment boundaries (in insect segments, for example), but bioelectric boundaries can form before any visible anatomical partitions do.

This is where the notion of a *bioelectric midline* becomes helpful. The embryonic midline is not just a geometric line down the center; it is a structure and a set of signals that keep left and right from bleeding into each other. In classic molecular accounts, the midline expresses inhibitors that prevent left-sided signals from spreading to the right. Bioelectric work suggests that an earlier midline-like barrier can exist in the form of a voltage discontinuity and a pattern of gap junction closure that restricts the movement of charged morphogens (patterning molecules) across the center. The embryo, in effect, can draw an invisible fence.

The fence metaphor matters because it offers a plausible way to get reliability out of small asymmetries. Suppose one side of the embryo becomes slightly more hyperpolarized than the other due to asymmetric placement of pumps such as the H^+-V-ATPase during early cleavages. On its own, that difference might be tiny, like a faint tilt in the terrain. But if the difference is used to bias the distribution of a determinant (say, serotonin or another small signaling factor), and if the embryo's connectivity network helps keep that determinant on one side, then the embryo can amplify a whisper into a decision.

This amplification also helps clarify why disturbing the early electrical step can randomize organ placement without necessarily destroying organ formation. Once the left side has been chosen, a robust gene network can take over, executing a program of asymmetric morphogenesis. Nodal signaling on the left, for example, is part of such a program in many vertebrates. Robust programs are good at making organs; they are not good at deciding which side is left. If the "side choice" is randomized early, the later program can still run, but it runs on a shuffled coordinate system.

There is a temptation to imagine a single master switch: an initial voltage difference that flips a genetic lever, and everything else follows. Real embryos are more like distributed systems. Multiple cues overlap in time. In some species, cilia-driven flow is central; in others, earlier intracellular or tissue-level asymmetries seem to play a larger role. Even within the same organism, the early bioelectric and chemical biases may not *replace* cilia and fluid flow; they may help establish the conditions that make flow interpretable, or they may provide parallel inputs that converge on the same downstream gene networks. Daniel Grimes and Rebecca Burdine (2017) emphasized how leftward flow and the Nodal—Pitx2 cascade coordinate asymmetric morphogenesis, while other work has argued that the links between the earliest physiological gradients and later asymmetric gene expression remain partly unresolved. That gap is not an embarrassment; it is exactly where active science lives. One can know that a stage is essential—and be able to disrupt it, rescue it, or reroute it—while still working out the intermediate steps.

A historical anecdote makes the conceptual tension clear. For much of the twentieth century, developmental biology was shaped by a preference for mechanisms that could be seen. Gradients of chemicals diffusing across tissues were easy to imagine; the movement of a fluid driven by cilia could be filmed. Electricity, by contrast, was either the province of physiologists studying nerves or a vague notion of "vital forces" best left behind. Even when researchers measured endogenous electrical fields around wounds or growing tissues, those measurements could be dismissed as byproducts, not instructions. The turning point was methodological as much as conceptual: once voltage-sensitive dyes, ion-selective electrodes, and genetic tools made it possible to map and perturb voltage patterns with spatial precision, bioelectricity became experimentally testable as a source of pattern, not merely a consequence.

The left—right problem is a particularly sharp test because the outcome is binary and the readout is unambiguous. A heart on the left is not a subtle phenotype. If manipulating an ion pump at the 4-cell stage makes that heart reliably appear on the right in a significant fraction of embryos, while leaving much else intact, then the embryo has just confessed that electricity participates in deciding sidedness.

The human stakes sit close to the surface. Left—right disorders are often discovered in the neonatal intensive care unit, not the anatomy lab. A baby with heterotaxia may have a complex congenital heart defect because heart looping is intimately tied to left—right cues; surgical decisions depend on the arrangement of major vessels and chambers. Other complications can involve the spleen (absent, multiple, or misplaced), the liver, or the intestines. Genetic causes exist, including mutations affecting cilia and signaling pathways, but environmental and stochastic factors also contribute. The bioelectric perspective does not magically provide a single diagnostic test. It does, however, broaden the set of plausible mechanisms and raises a provocative possibility: that some cases might originate from very early physiological disruptions, perhaps transient, long before any prenatal scan could detect them.

It also reframes what it means for the genome to "control" development. Genes encode ion channels and pumps. Those proteins, when placed in the membrane, generate V_{mem} patterns. Those patterns can bias the distribution of small molecules and the activity of voltage-sensitive biochemical pathways, ultimately influencing gene expression again. Left—right asymmetry, in this light, is not purely genetic, nor purely mechanical. It is a circuit in which electrical state is a real variable, one that can store information about position and propagate it across cells faster than diffusion alone.

A final subtlety is worth lingering on because it rescues the story from sounding mystical. Voltage is not a message floating free of matter. It is a bookkeeping device for charge separation, and charge separation is inseparable from ion movement, and ion movement is inseparable from proteins made by genes and arranged by cell biology. The novelty is not that embryos "use electricity" as a magical force; it is that the embryo exploits the fact that voltage is a *collective property*. A single cell's membrane potential is already an integration of many molecular events. When cells couple through gap junctions, their voltages become a tissue-level conversation. Left and right can differ not because any one gene is "left" and another is "right," but because the embryonic tissue, as an electrical network, falls into a slightly asymmetric state and then locks that state in by controlling what can cross the midline.

The embryo's first left—right decision is therefore not a miniature version of adult anatomy. It is an invisible asymmetry in the flow of ions and small molecules: protons pumped, channels opened, junctions gated, charged determinants shepherded to one side. Later, when the heart tube loops and the gut rotates, those dramatic motions are the visible handwriting of an earlier electrical sketch. The startling part is not that life depends on chemistry—it always has—but that, at the very beginning, chemistry is guided by something as spare and physical as a voltage difference, a faint spark that tells an otherwise symmetric body how to be reliably, functionally lopsided.

3.2 The Electric Face

There is a moment in every vertebrate life when the future face is, quite literally, nowhere to be found. No eyes, no jaw, no nostrils; not even the hint of a mouth. Yet the embryo is not improvising. Cells are already being assigned roles, given destinations, and nudged toward fates that will later look inevitable. The surprise is that one of the earliest reliable previews of where a face will form is not a ridge, a fold, or a recognizable patch of tissue, but a pattern of voltage—a faint electrical "mask" draped over the embryo long before anatomy catches up.

The idea sounds implausible until one remembers what voltage means at this scale. A cell membrane is a thin electrical barrier. Ion channels and pumps move charged particles across it, and the imbalance creates a membrane potential, V_{mem}. A single cell's V_{mem} is not a metaphor; it is a measurable number. In an embryo, the more striking fact is that V_{mem} is not uniform. Groups of cells can share a voltage state, and the boundaries between those states can be sharp. The pattern is not a decorative glow on a microscope image; it is a living variable that cells can sense, exchange through gap junction connections, and translate into biochemical decisions. Earlier, left—right patterning made use of tiny asymmetries in this electrical landscape. In the head, the landscape becomes more elaborate: not a single tilted plane, but a map.

A face is not a single organ. It is a negotiated settlement among many cell populations, each arriving with its own history. Some facial tissues come from the surface ectoderm (the outer layer that also makes skin). Much of the cartilage, bone, and connective tissue of the face comes from *neural crest* cells, a remarkable population that begins near the developing nervous system and then migrates like a diaspora, streaming into the front of the embryo and fanning out to build the jaw, the middle ear, the palate, and much else. Neural crest cells are famously plastic: the same kind of cell can become pigment, peripheral neurons, facial cartilage, depending on the signals it encounters. That plasticity is a strength, but it is also a risk. A face made from highly migratory, highly flexible cells needs good guidance and good timing. The embryo cannot simply tell them "be face"; it must tell them where to go, when to stop, and what geometry to respect.

If one wanted a mechanism that could guide such a process, one might choose something continuous and spatial: a gradient, a boundary, a set of coordinates that cells can read as they move. Developmental biologists have long known chemical gradients that do this job. The unexpected addition is that voltage gradients can do it too, and sometimes earlier than the chemical markers we are used to thinking of as the starting gun.

A concrete entry into this discovery came from work in Michael Levin's laboratory, particularly by Vaibhav Pai and colleagues (2012), using the frog *Xenopus laevis*. Frogs are not chosen for cuteness, though they have that going for them. They are chosen because their embryos develop outside the mother, are robust, and can be manipulated with a precision that would be impossible in many mammals. One can inject RNA to misexpress an ion channel in a particular lineage, or apply a drug at a specific developmental window, and then watch what happens to organs that form much later.

Pai and collaborators used voltage-sensitive dyes to visualize V_{mem} patterns in the developing embryo. Instead of waiting for the eye to become visible as a bulge or a pigmented structure, they looked earlier, at the stage when the anterior neural plate (the tissue that will give rise to brain and retina) is present but still flat and featureless. What appeared was a distinct cluster of hyperpolarized cells—cells whose interiors were more negative than their neighbors—in the region that would later contribute to the eye. It was a physiological signature of the eye field, arriving before the physical eye primordium could be pointed out to the naked eye.

The story would be merely decorative if the voltage pattern were just correlated with later anatomy. Correlation, in embryos, is cheap: almost everything correlates with everything else if you look at the right time. The deeper question is causality. Does the hyperpolarized region help *cause* the eye, or is it a side effect of the same genetic program that will later build an eye anyway?

The team addressed this with the kind of experimental move that makes developmental biology feel like science fiction. They deliberately altered membrane potential in targeted cell populations. When they depolarized lineages that normally contribute to the eye, eye development was disrupted. More strikingly, when they changed V_{mem} in cells that were *not* fated to become part of the eye, they could induce ectopic eyes—additional eyes in unusual places, sometimes far from the head. These were not vague patches of pigment. They displayed recognizable anatomical and histological features of eyes. The details matter because they shift the interpretation. If changing voltage in a distant patch of tissue can recruit the machinery needed to make a complex organ like an eye, then V_{mem} is acting less like a permissive background condition and more like an instructive signal: a message that can be "read" by the developmental system and translated into a plan.

This idea is unsettling for an intuitive reason. Many people imagine the genome as a blueprint: genes specify what structures appear and where, and the embryo simply executes. But a true blueprint has coordinates written into it. DNA is not arranged as a map of the body; it is a list of parts and rules. Voltage patterns, by contrast, can *be* coordinates in a literal spatial sense. They are patterns across a tissue, and tissues are what bodies are made of.

A helpful analogy is a faint pencil sketch beneath a painting. The artist may later lay down thick pigment, add shading, and create a lifelike face, but the early pencil marks constrain where eyes and mouth can plausibly go. They are not the final image, but they prevent the final image from drifting into chaos. In the embryo, the "pencil marks" are not a conscious plan, and they are not static. They are dynamic fields that can be strengthened, erased, and rewritten by physiological changes. The point is that they exist early enough to coordinate migration and differentiation before those processes become irreversible.

The phrase "electric face" risks sounding like a metaphor. It helps to be concrete about what cells can do with voltage. A change in V_{mem} can alter the entry of calcium through voltage-gated channels; calcium then activates enzymes and transcription factors that change gene expression. Voltage can drive the transport of charged signaling molecules through membrane transporters. It can influence the opening probability of channels that regulate cell volume, and cell volume affects mechanical properties and movement. It can modulate the distribution of small molecules shared through gap junctions. None of these routes require mysticism, only the recognition that cells are electrochemical machines.

In the developing head, small differences can have outsize consequences because craniofacial development is an exercise in choreography. Neural crest cells emerge, migrate, pause, and condense into arches and prominences. Those prominences then grow and fuse with exquisite timing; the palate, for example, must elevate and meet at the midline. Minor shifts in where a cell population condenses can lead to major changes in facial proportion. That is one reason craniofacial anomalies are among the most common congenital differences in humans: the system is complex, and complexity invites fragility.

This fragility becomes painfully visible in real disorders, and here an electrical perspective adds explanatory power. Andersen—Tawil syndrome is a human channelopathy, a disease caused by mutations in an ion channel gene, KCNJ2, which encodes the potassium channel Kir2.1. Patients can have periodic paralysis and cardiac rhythm abnormalities; some also display characteristic craniofacial features. Dany Spencer Adams and collaborators (2016) explored the developmental side of this channelopathy using *Xenopus* and mouse, asking whether misregulating such a channel could perturb craniofacial patterning through changes in V_{mem}.

The key finding was that misexpression of Kir2.1 variants altered the spatial pattern of resting membrane potential in the ectoderm of neurulating embryos, and that this disruption was accompanied by shifts in expression of craniofacial regulatory genes, including markers for cranial neural crest and placodes. The result is important not because one frog embryo perfectly models a human syndrome, but because it supports a more general claim: many distinct ion translocators can produce similar craniofacial defects *if* they perturb the voltage pattern. That suggests that the relevant variable is not the identity of a particular channel as a "craniofacial gene" in the narrow sense, but the bioelectric state that the channel helps create.

A historical anecdote makes this shift feel less abrupt. In the nineteenth century, long before genes were discovered, physicians recognized that certain congenital malformations clustered in families. Later, as genetics matured, these clusters were reinterpreted as evidence of mutations in specific genes. The gene-centered picture was enormously successful, but it also encouraged a subtle habit: to treat each gene as if it were a miniature organ blueprint. Channelopathies complicate that habit. An ion channel gene is not "for" a face any more than it is "for" a heart rhythm. It sets electrical conditions, and those conditions are relevant wherever timing and pattern are sensitive to voltage. The same mutation can therefore produce symptoms in muscle, heart, and craniofacial development, not because the gene has three unrelated jobs, but because voltage is a shared control variable.

There is another twist, more conceptually delicious, that comes from laterality. Eyes are paired structures, and in a purely symmetric world, one might expect the electrical preparation of the left eye and right eye to be identical. Yet Pai, Vandenberg, and colleagues (2012) reported a consistent bioelectrical left—right asymmetry in neurally derived tissues in *Xenopus*. In particular, the hyperpolarization associated with eye induction often appeared first in the right eye field more often than in the left. The difference is subtle—the embryo still makes two eyes—but the subtlety is the point. It suggests that the left—right coordinate system established earlier can leave fingerprints even on paired organs that end up looking symmetric.

Why should anyone care which eye field hyperpolarizes first? Because it shows that laterality is not merely about placing the heart on the left. It is a global property of the embryo's physiology, one that can bias timing, sensitivity to perturbation, and perhaps even vulnerability to defects. Indeed, perturbations that disrupt visceral left—right patterning can also randomize this eye-field bioelectrical asymmetry, hinting that the same upstream mechanisms that decide organ placement may tune the head's electrical map as well. Development, in other words, does not solve left—right once and then forget it; the solution seeps into multiple tissues as a background condition, like a slight but consistent wind that affects many moving parts.

At this point the "electric face" is not only a claim about eyes. It is a model of how craniofacial features are prefigured: by physiological domains that appear early, guide migration and gene expression, and can be experimentally rewritten. The most compelling evidence comes from the combination of prediction and intervention. Voltage patterns can predict where an organ field is forming, and interventions that shift those patterns can shift the outcome. That is the signature of a true control layer.

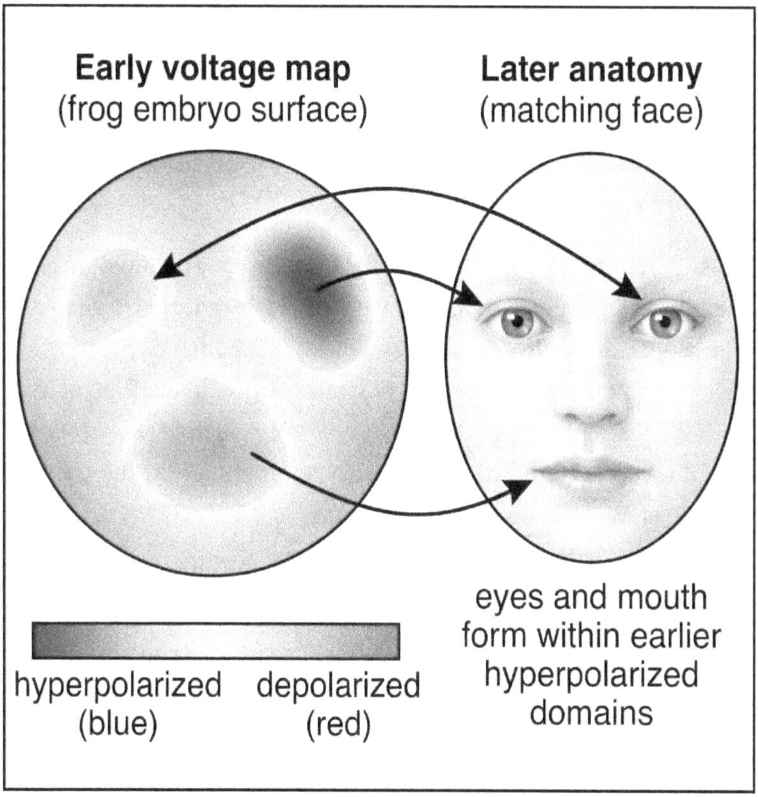

The figure is deliberately schematic: real embryos are not perfect ovals, real voltage maps are mottled and dynamic, and the relationship between an early electrical domain and a later organ is not a one-to-one stamping of shape. The point is the logic. A hyperpolarized region appears first, marking a location; later, an anatomical feature arises in that location. Change the region, and the feature can shift or multiply.

What makes this especially compelling is that craniofacial patterning is not merely about placing parts, but about ensuring compatibility. Eyes must connect to the brain; jaws must align with muscles and nerves; a mouth must open into a digestive tract. Ectopic eyes, when induced experimentally, raise a profound question: how much of that compatibility is preprogrammed in a local tissue, and how much is negotiated by the developing system as a whole? The answer appears to be: more negotiation than one might expect. A voltage change can recruit a multi-step developmental cascade that includes both local differentiation and integration into a larger plan, at least to a surprising degree.

This shifts the emotional tone of the science. The embryo is not merely a fragile object that fails when perturbed; it is also adaptable. The same adaptability that makes congenital anomalies possible also makes repair and compensation possible. A system that can interpret an electrical cue and build an organ is, in principle, a system that might be coaxed toward regeneration or correction, if the cues can be delivered safely and at the right time.

Even without thinking about therapies, the "electric face" changes how one looks at a developing head. Instead of searching only for genes turning on in a particular patch of tissue, one can also look for physiological states: zones of hyperpolarization and depolarization, corridors of gap junction connectivity, and time windows when a temporary electrical map is doing real work. In many embryos, the face does not emerge as a single structure, but as a convergence. Neural crest streams arrive; placodes thicken; epithelia bend and fuse. If a voltage map is present early, it can serve as a shared reference, a way of making sure that independent cell populations agree on where "here" is.

It is hard to watch a newborn focus on a parent's face without feeling the weight of what had to go right. Eyes must be placed with a precision that allows binocular vision; jaws must fit; the palate must seal. These are not merely aesthetic features. They are prerequisites for feeding, breathing, sensing, and communicating. That such a complex structure can be foreshadowed by a faint, shifting electrical pattern on the surface of an embryo is one of those facts that leaves the world feeling slightly stranger, but also slightly more intelligible: form is not only built from molecules and mechanics, but also from fields that organize them, quietly, before there is anything recognizable to organize.

3.3 Software Controls Hardware

A genetic explanation of development is comforting because it seems to offer solid ground: DNA is stable, measurable, and written in a language we can translate into proteins. But embryos do not grow by reading DNA the way a printer reads a file. Genes are not little sculptors. They are instructions for making molecules, and those molecules must organize themselves across space and time into a body. Somewhere between "a cell makes a protein" and "a face has two eyes in the right place" lies a gap that cannot be bridged by naming more genes. The missing ingredient is not more parts, but a way to coordinate parts.

Bioelectricity offers a candidate for that coordination layer, precisely because it sits at an awkward but powerful intersection. Ion channels and pumps are proteins encoded by genes. Their combined activity sets a cell's membrane potential, V_{mem}, and that electrical state can spread through tissues as cells connect via gap junctions. Then, crucially, the electrical state feeds back: it changes which genes are expressed, which proteins get activated, where small molecules accumulate, how cells migrate, and what shapes tissues can plausibly form. The genome builds the electrical circuitry; the electrical circuitry, in turn, helps decide how the genome is read.

It helps to borrow a distinction from computing, with the caveat that no embryo is a laptop. In a computer, hardware is the physical machinery and software is the pattern of states that tells that machinery what to do. A computer's software is not a separate ghost; it is a real configuration of voltages in circuits. Yet it has properties that feel different from the hardware: it can be updated, copied, debugged, and sometimes it can produce similar outputs even when running on different machines. In development, ion channels, pumps, and gap junction proteins are hardware. Tissue-scale patterns of V_{mem} are closer to software: dynamic states that can be stable enough to guide large-scale structure, yet flexible enough to be altered by context.

That analogy becomes more than a rhetorical flourish when one notices what voltage patterns can do. A chemical gradient can inform a cell about position by concentration. A voltage pattern can inform a cell about position by electrical state, which is a compact summary of many variables at once: extracellular ions, intracellular ions, channel expression, pump activity, metabolic energy, connectivity to neighbors. V_{mem} is not simply another molecule; it is a state variable, a kind of physiological "vote" that many molecular processes cast ballots into. That is why voltage is so useful for coordination, and also why it is hard to reduce to a single pathway.

The bidirectionality matters. If genes simply determined voltage and voltage simply acted downstream, bioelectricity would be another branch of genetics. The interesting claim is that voltage and gene expression form a loop. Loops have consequences. A loop can amplify a weak signal, stabilize a pattern against noise, and create memory. In an embryo, memory is not a quaint concept; it is the ability of a tissue to hold on to a positional decision long enough for cells to divide, move, and differentiate without losing track of what they are collectively building.

Consider a cell in an early embryo deciding whether it belongs to an eye field, a jaw field, or a patch of ordinary skin. Genes contribute by providing transcription factors and receptors that define competence. But competence is not destiny. If the local V_{mem} is hyperpolarized in a characteristic way, that state can bias calcium dynamics, transporter activity, and second messenger systems that activate certain transcription factors and silence others. In the frog eye work by Pai and colleagues (2012), changing V_{mem} could redirect fate strongly enough to produce ectopic eyes. The intervention was not a new gene network invented from scratch; it was a shift in a physiological variable that pushed existing networks into a different attractor state, to borrow the language of dynamical systems.

What do cells use to translate voltage into gene expression? The embryo does not have a single "voltage sensor gene" any more than a city has a single "traffic sensor." It has many mechanisms that are sensitive to electrical state. Voltage-gated calcium channels open when the membrane depolarizes past a threshold, allowing calcium ions to enter. Calcium then binds to proteins like calmodulin and activates kinases and phosphatases that modify transcription factors. Some transporters move signaling molecules in a way that depends on membrane potential; if the electrical gradient changes, so does the movement of molecules that can enter the nucleus and influence transcription. There are also enzymes whose activity depends on voltage-sensitive lipid signaling at the membrane. Different tissues may rely on different mixtures of these decoders, which is one reason a single electrical perturbation can have distinct outcomes depending on timing and location.

Timing is not a minor detail; it is the difference between a nudge and a catastrophe. A voltage change that is harmless in an adult tissue can be transformative in an embryo if it arrives during a short window when pattern is being set. Craniofacial development illustrates this sharply. In work connecting Kir2.1 (KCNJ2) to craniofacial morphogenesis, Adams and colleagues (2016) found that altering V_{mem} patterns during neurulation shifted the expression domains of craniofacial regulatory genes and led to facial anomalies. The most conceptually important part of that work was not merely that one potassium channel mattered, but that multiple different ion translocators could produce similar phenotypes as long as they produced a similar electrical change. That is exactly what one would expect if the developmental system cares about the *state* more than the *component*. In software terms, the program is written in voltages, not in the brand name of the transistor.

A historical anecdote helps clarify why this is still a conceptual hurdle. For decades, biologists were trained to treat electrical phenomena as outputs of excitable tissues: nerves and muscles. When early twentieth-century researchers reported electrical currents in developing tissues and regenerating limbs, the observations were often filed under curiosities. They were hard to connect to the reigning explanatory framework, which centered on genes and morphogens. Michael Levin's synthesis in the 2010s, including his reviews and the later "bioelectric code" concept developed with Christopher Martyniuk (2017), did not introduce electricity into biology so much as insist that developmental biology take seriously what electrophysiologists had always known: bioelectricity is not limited to rapid spikes. Slow, steady membrane potentials and gap junction networks can process information.

The phrase "process information" can provoke skepticism, so it is worth being precise. Information here does not mean consciousness. It means that a pattern can represent a constraint on future behavior. If a group of cells is coupled by gap junctions, they can share ions and small molecules, and their V_{mem} values can influence each other. Such a network can settle into stable patterns, the way a row of coupled pendulums can synchronize. If those patterns correspond to anatomical decisions—"this is where an eye field should stabilize," "this is where a boundary should hold"—then the network has a kind of memory. It will resist perturbations, and it will tend to restore the pattern after small disruptions. That is precisely the kind of stability development needs.

One can see the loop more clearly by walking around it step by step.

Genes encode ion channels and pumps. These are proteins that either allow specific ions to pass (channels) or actively move ions using energy (pumps). A cell's membrane potential, V_{mem}, emerges from the combined activity of many such proteins, plus the concentrations of ions inside and outside the cell. Gap junction proteins, also encoded by genes, connect neighboring cells and allow electrical and chemical coupling.

Now V_{mem} is not just a number on a voltmeter. It changes the probability that certain channels open, changes calcium entry, and changes the operation of transporters. Calcium and other second messengers activate signaling pathways that reach the nucleus, turning transcription factors on or off. Those transcription factors change gene expression, including the expression of the very ion channels, pumps, and gap junction proteins that set V_{mem} in the first place. The circle closes.

When a loop closes, a new possibility appears: the system can have more than one stable state. A cell population can be pushed into one state or another depending on initial conditions and signals. That helps make sense of developmental plasticity. It also helps explain why certain interventions, applied briefly, can have long-lasting anatomical consequences. A transient voltage shift can tip the system into a different stable regime of gene expression, after which the new regime maintains itself.

This perspective also casts a different light on a classic developmental question: how does an embryo achieve scale and proportion? A chemical morphogen gradient can, in principle, scale with tissue size, but doing so robustly is not trivial. Voltage patterns, because they are network properties, can scale in different ways. A coupled tissue can adjust its electrical domains as it grows, maintaining boundaries and relative sizes through feedback. This is not guaranteed, and it is not magic; it is the ordinary behavior of coupled dynamical systems. The biological point is that tissues are full of feedback loops, and bioelectric feedback is a particularly fast and integrative one.

There is a practical reason biologists care about this beyond intellectual satisfaction. If voltage patterns are part of the control layer, then some developmental disorders, and perhaps some diseases of adulthood, might be understood as misregulation of that layer. Cancer, for instance, has been described by some researchers as involving altered bioelectric states, with depolarization associated with proliferative, less differentiated behavior. Even when one sets cancer aside, the general idea matters: a channel mutation or an environmental toxin that subtly shifts ion flux could have effects that look "genetic" (altered gene expression, altered pattern) without changing DNA sequence at all. That is a form of epigenetic control in the broad sense: a stable change in gene expression driven by state, not sequence.

The open question, and the source of healthy skepticism, is predictability. It is one thing to say that changing V_{mem} can change anatomy; it is another to claim that one can read a voltage map and predict, in detail, the face that will result. Development is not a simple codebook, where "-60 mV means eye" and "-30 mV means mouth." Voltage patterns are interpreted in context: by the tissue's gene expression, by its connectivity, by its mechanical environment, and by its history. Different species may use similar electrical components but wire them differently in time. The fact that some mechanistic links remain unknown is not a weakness of the idea so much as a sign that it is a real scientific hypothesis: it generates specific questions, and it can be wrong in informative ways.

A concrete way to think about this is to ask what would count as a mature "bioelectric software" understanding. It would not be a single universal voltage map that fits all embryos. It would be the ability to specify, for a given tissue at a given time, what pattern of V_{mem} is stable, how it can be perturbed, what downstream transcriptional changes it will tend to produce, and how the tissue will respond if part of the pattern is damaged or rewritten. That is an engineering-style description: not merely naming parts, but understanding control.

The appeal of this view is that it makes room for both genes and physiology without forcing them into a hierarchy. Genes matter because they build the machinery. Physiology matters because it is the state of the machinery, and states can do causal work. Once this is granted, many puzzling results become less puzzling. Why can two different channel mutations lead to similar craniofacial defects? Because they converge on a similar V_{mem} disturbance. Why can a brief perturbation early in development lead to a stable change much later? Because it flips a feedback loop into a different regime. Why can a tissue sometimes regenerate a missing structure with the right proportion? Because it may be maintaining a pattern goal in a distributed physiological state, not merely executing a one-way genetic script.

It is tempting to end with a grand claim that electricity is the hidden master of form. That would be as misleading as the old belief that DNA is the sole master. Embryos are not ruled by one cause. They are managed by many interacting layers: gene regulation, chemistry, mechanics, and bioelectricity among them. What the bioelectric view adds is a sense of *how* a body can be coordinated at all: by using electrical states as a shared medium that can couple cells into tissues, tissues into regions, and regions into an integrated plan that remains coherent even as everything is dividing, moving, and changing its identity. The embryo builds its own control system as it goes, and then lives inside it.

Chapter 4

Recall of the Flesh

Cut a flatworm into a hundred pieces, and it will rebuild a hundred perfect bodies. We explore the "electric face" of an organism-a detectable voltage scaffold that instructs cells on where to build and when to stop, offering a radical new path for inducing regeneration in humans by rewriting the memory of the flesh.

4.1 The Immortal Worm

A planarian looks like something a child might sketch after peering into a jar: a small, flat ribbon of life, arrow-headed, gliding on invisible cilia. It is easy to underestimate. Then someone cuts it.

Not cruelly, in the laboratory sense: a clean slice with a scalpel under a microscope, sometimes into three pieces, sometimes into ten, sometimes into fragments so small they scarcely seem like animals at all. Days later, what had been a headless scrap has grown a head. A tail-less scrap has grown a tail. With enough patience, a whole cohort of complete worms swims where a single worm once did.

The first temptation is to tell a familiar story: genes as blueprints, stem cells as builders, and regeneration as a local construction project triggered by injury. Planarians force a stranger conclusion. They do have genes, and they do use stem cells—remarkably powerful ones. But the striking fact is not merely that they can regrow *the right shape*, to the right proportions, in the right orientation, and they stop when the job is done. That kind of success implies a reference: a stored pattern against which the rebuilding is measured.

A historical anecdote shows how long this has been unsettling. Thomas Hunt Morgan—later famous for fruit fly genetics—spent part of the 1890s and early 1900s slicing planarians and thinking hard about what he saw. He was fascinated by the way a fragment could reorganize itself into a proportional animal rather than a crude patchwork. Even before anyone knew about DNA, planarians made it hard to believe that form was simply "in the parts." Something global seemed to supervise the local efforts, as if the fragment knew not just how to make tissue, but what the finished animal ought to look like.

Modern molecular biology filled in part of the picture. Planarians carry a population of adult stem cells called *neoblasts*. When the worm is wounded, neoblasts proliferate and form a blastema: a mound of new tissue at the cut edge. This is the growth side of regeneration. But there is also *morphallaxis*: the reshaping of existing tissues so that old and new match, scaling and proportion are restored, and organs land in sensible places rather than merely appearing somewhere. Reddien and Sánchez Alvarado (2004) captured this dual nature clearly: regeneration is not simply replacing missing parts; it is rebuilding an integrated body plan.

Even the neoblast story, though, points beyond the idea of a purely genetic blueprint. Wagner, Wang, and colleagues (2011) showed that some neoblasts are *pluripotent* in an unusually strong sense: a single transplanted cell can rescue an irradiated planarian and repopulate the tissues. That is an astonishing cellular capacity. Yet the presence of powerful stem cells does not explain how a fragment decides *what* to build at each end. Potent builders are not the same as an architect. The puzzle of architecture is what makes planarians central to bioelectric thinking.

One reason planarians acquired an aura of immortality is that many lineages can be maintained by fission and regeneration for long periods, as if the body continually resets itself. Telomeres—the protective ends of chromosomes that tend to shorten with cell division—are part of the mechanistic folklore here. Tan, Rahman, and colleagues (2012) reported that asexual *Schmidtea mediterranea* maintain telomere length across repeated fission and regeneration, consistent with telomerase activity in somatic tissues during regenerative episodes. The details are nuanced (sexual and asexual animals behave differently, and "immortal" is a dangerous word in biology), but the impression is still instructive: a planarian treats damage not as an endpoint but as an opportunity to restore its internal bookkeeping.

The deeper lesson is not about lifespan. It is about memory. When a planarian repairs itself, it behaves as if it remembers a target shape. The memory is not neuronal in the ordinary sense. A fragment without a brain does not merely survive; it makes a new brain in the correct place. The memory is not written directly in the immediate genome either, because all fragments share essentially the same DNA, yet they produce different structures depending on their position along the head-to-tail axis and on the signals they receive after cutting. The same genome can support head-making and tail-making. Something else decides which program is executed at a given wound.

Bioelectricity offers a candidate for that "something else": large-scale patterns of voltage across tissues, and the networks of cell-to-cell electrical communication that stabilize those patterns.

Every living cell maintains a voltage difference across its membrane, the *membrane potential*. This voltage is created by ion channels and pumps that move charged atoms (ions such as potassium, sodium, chloride, and protons) across the membrane. A single cell's voltage can influence its behavior: whether it divides, migrates, differentiates, or remains quiescent. In a tissue, cells are often connected by *gap junctions*—protein channels that allow ions and small molecules to pass directly between neighboring cells. Gap junctions turn a sheet of cells into an electrically coupled network. Instead of each cell being an isolated battery, the tissue can act like a distributed circuit.

In Michael Levin's lab, a key idea emerged in the 2000s and 2010s: the pattern of membrane potentials across the planarian body is not just a readout of development; it can be an *instructive* signal that helps set the body plan. The phrase "bioelectric code" is provocative, but it captures something simple: spatial voltage patterns can carry information. Not information in the digital sense of bits on a hard drive, but information in the physiological sense—constraints and biases that push living matter toward one anatomical outcome rather than another.

Planarians make this visible because their outcomes can be spectacularly wrong in a way that still looks organized. A worm can be induced to regenerate with two heads: one at the usual front end and one where the tail should be. This is not a random deformity. It is a coherent re-specification of polarity: tail becomes head.

One route to two-headedness came from interfering with gap junctions. Oviedo, Morokuma, and colleagues (2009) showed that blocking gap junctional communication during a critical window after amputation can yield bipolar worms. The timing matters: there is an early decision period when the wound interprets its position in the whole-body context, and when long-range cues can be disrupted. This fits a circuit intuition. If the tissue's electrical network normally integrates information across distances, then disconnecting the network at the wrong time might prevent the wound from "learning" that it is posterior. Deprived of the global context, the local default could drift toward head formation, or the posterior identity signal might fail to arrive.

Another route targets the cellular hardware that establishes voltage in the first place. Beane, Morokuma, and colleagues (2011) used a chemical genetics approach to show that a proton/potassium pump (H,K-ATPase) is required for proper head regeneration. When this pump is inhibited, the anterior blastema fails to acquire its normal depolarized state, and head regeneration can be blocked even though a blastema forms. The animal is attempting to regenerate, but the identity of the new tissue is mis-specified. The implication is sharp: voltage is not merely correlated with "headness"; it participates causally in the decision.

Once one accepts that bioelectric states can influence polarity, a more unsettling possibility follows: perhaps the "target morphology"—the specification of what counts as a correct body plan—is stored, at least partly, in stable patterns of electrical connectivity and resting potential. That would mean the worm's body plan is not only encoded in DNA and molecular gradients, but also in the dynamical state of a physiological network. If so, then altering the network could rewrite what the animal rebuilds, even when the genome is unchanged.

This is where planarians begin to resemble not just regenerating tissues, but reprogrammable systems.

Durant, Morokuma, and colleagues (2017) pushed this idea further by showing that brief perturbations of bioelectric networks can lead to long-lasting changes in regenerative outcomes. In some cases, a transient treatment produces worms that look normal but later regenerate abnormally when cut again—a kind of cryptic rewrite. The animal carries an altered "set point" for patterning that is not obvious until regeneration tests it. That is precisely what one expects from a distributed circuit with multiple stable states: the system can be nudged into a different attractor, and thereafter it behaves differently even after the original perturbation is gone.

The two-headed worm is the simplest way to grasp the concept, because it compresses a complex argument into a visual punchline. Two heads mean two anterior organizers, two sets of sensory structures, two mouths in some species, and a topology that is coherent even while it violates the species' usual blueprint. The genome has not been edited. The stem cells are the same. The injury is the same scalpel stroke. The difference is the physiological state of the tissue network that interprets the injury.

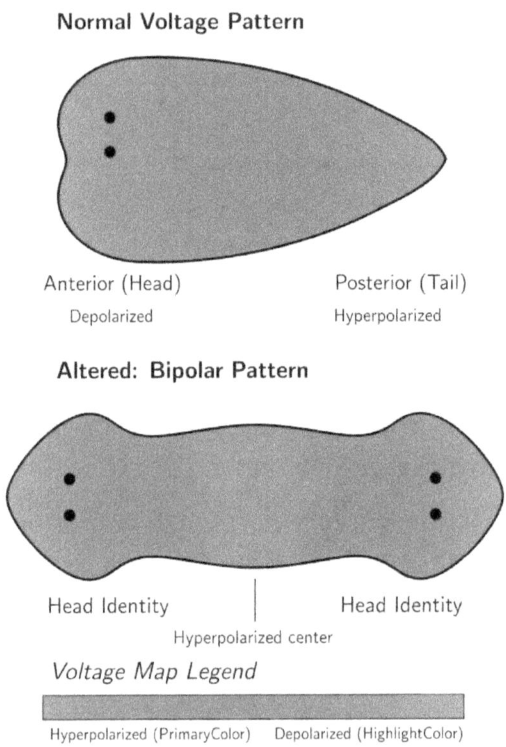

It is worth lingering on what "stored in circuits" means, because it can sound mystical if taken the wrong way. No one needs to imagine miniature lightning bolts sketching organs in the air. The proposal is more sober: bioelectric states are *control variables*. They bias gene expression, protein localization, and signaling pathways, partly through well-known intermediates like calcium influx, voltage-sensitive transport of small molecules, and the regulation of transcription factors. In other words, voltage patterns sit upstream of molecular cascades that ultimately build tissue. A change in voltage can be translated into a change in chemistry, and chemistry changes what cells do.

Why call this a memory? Because the state can persist and because it can influence future outcomes. A planarian's epithelium and muscle are not passive scaffolds; they are living networks that can maintain spatial differences in resting potential. Gap junction coupling can stabilize these differences over distances, like a fabric of communication. A wound then becomes not merely a tear that triggers growth, but an input into a system already holding a pattern. The same injury, applied to tissues in different network states, can lead to different anatomical decisions.

This helps explain a subtle feature of planarian regeneration that genetics alone does not make intuitive. Planarians restore proportion. Remove a small amount, and the animal repairs with minimal reshaping. Remove the head, and the fragment not only makes a new head; it also remodels itself so that the head is not comically large or small relative to the remaining body. This suggests that the animal is regulating toward a global set point: the "correct" planarian is not just a list of parts, but a geometry.

Bioelectric networks are well-suited to this kind of regulation because they naturally implement integration and feedback. A single cell's voltage is local, but a tissue's voltage pattern can be global, because the tissue literally allows charges and small messengers to spread. If one region of the worm is strongly depolarized and another is strongly hyperpolarized, that spatial difference can serve as a coordinate system. Cells do not need to "know" their absolute position by reading a ruler; they can read their position relative to a physiological gradient.

The two-headed experiments sharpen the point: the coordinate system can be edited. One can nudge the network so that the posterior end behaves as if it occupies an anterior coordinate. When the animal is cut, the stump consults its coordinate, and the reconstruction follows suit.

There is an unavoidable tension here with the way many of us have been trained to think. We are comfortable with the idea that the genome encodes proteins, and proteins make structures. But planarians demonstrate that the same molecular toolkit can be orchestrated into different macroscopic outcomes depending on high-level physiological context. The genome is necessary—without it there is no worm, no ion channels, no gap junctions, no neoblasts—but it is not sufficient to specify, at every moment, the correct anatomy. The living system must also store state in how its parts are connected and in the electrical conditions those connections maintain.

One can make an analogy to a city after an earthquake. The bricks are the same bricks. The workers are the same workers. The building codes are the same documents. Yet rebuilding can either restore a street layout or produce a confused sprawl, depending on whether there is a functioning map, a chain of command, and a shared sense of what "the city" is supposed to be. In a planarian, the map is not printed on paper. It is embodied in physiology: in coupled cells holding voltage differences that, in turn, constrain what patterns of growth and remodeling are permitted.

This also clarifies why "immortal" is, in practice, a claim about *organization* as much as about cells. Regeneration is not merely the replacement of worn components; it is the maintenance of a pattern over time, despite cuts, despite missing parts, despite the constant turnover of molecules. The worm's persistence is the persistence of an attractor: a stable form toward which the system returns.

That is why the planarian is such a powerful ambassador for bioelectric morphogenesis. It is not the fastest regenerator, nor the most charismatic, nor the closest to humans. It is simply honest about what is happening. Slice it, and it reveals that living matter can carry a memory of shape that is not reducible to DNA sequence. It is written, at least in part, in the quiet voltages that cells maintain from moment to moment, and in the circuits they form together when they decide, collectively, what counts as a whole animal.

4.2 The Current of Injury

A cut is not only a tear in flesh. It is a break in an electrical surface.

Skin and many other epithelia (the tight, protective sheets lining organs and body surfaces) behave like living batteries. Cells in an epithelium spend energy to keep ions unevenly distributed: sodium tends to be handled differently on the outside than on the inside, chloride is actively moved, protons are pumped, potassium leaks and is reclaimed. Tight junctions between cells help maintain separation between compartments. The result is a voltage difference across the epithelial layer, called the *transepithelial potential*. It is not mystical. It is a measurable consequence of pumps, channels, and barriers doing their daily work.

Now imagine puncturing that battery. The moment you wound an epithelium, you create a low-resistance shortcut between regions that were carefully kept apart. Ions rush. Electric fields form. Currents flow around the wound edge, not because the body is trying to send a message in the way a nerve sends a message, but because physics insists on balancing charges when barriers are breached. Biology then takes hold of that physical event and shapes it. Channels redistribute, transporters change activity, calcium surges, gap junctional connections open and close. The injury current becomes not just a passive leak but a controlled signal with a characteristic time course: strong early, maintained for a while, then diminishing as the wound closes and the barrier is restored.

The term *current of injury* has a long and slightly tangled history. In the nineteenth century, physiologists studying excitable tissues such as muscle and nerve used it to describe electrical potentials associated with damaged tissue. In the twentieth century, it became part of the lore of regeneration research, especially through the work of Robert O. Becker, an orthopedic surgeon and researcher who took the idea seriously enough to carry electrodes to limb stumps. Becker's measurements and interpretations, reported across the 1960s and 1970s, helped persuade a generation that healing had an electrical dimension worth studying rather than dismissing as artifact. Even when later work refined the mechanisms, the core intuition survived: injury produces a distinct electrophysiological state, and that state can matter.

A difference that matters in practice is the difference between a wound that scars and a wound that regenerates.

Most adult mammals, including humans, heal by sealing quickly and then laying down collagen-rich scar tissue. This is lifesaving. A leaky body is a short-lived body, and evolution is ruthlessly pragmatic. But scarring comes with a price: function and form are only partly restored. A severed fingertip may regrow in a child under certain conditions, but an amputated limb does not. A deep skin wound can close beautifully and still erase hair follicles, sweat glands, and the original architecture. In the central nervous system, scarring can be especially devastating, walling off damage but also blocking regrowth.

By contrast, salamanders—the famous axolotl among them—can regrow an entire limb: bone, muscle, nerves, skin, blood vessels, and joints arranged correctly, and scaled to match the animal. The gap between these outcomes has often been framed in biochemical terms: growth factors, immune responses, gene expression programs. Those matter enormously. Yet there is also a physiological difference that appears early, before any obvious anatomy has changed: the electrical profile of the wound.

To see why this could be decisive, it helps to think about what a wounded stump needs to accomplish. It must recruit cells, coordinate their movements, decide which tissues to rebuild, and keep rebuilding until the original pattern is restored. Chemical gradients can do some of this. Mechanical cues can do some. But electric fields have a peculiar advantage: they provide a directional signal that cells can read at a distance without waiting for molecules to diffuse. An electric field is not a substance that must be produced and carried. It is a condition of space, created instantly when charges are separated. In a wet, salty environment like tissue, fields decay with distance and are shaped by conductivities, but over the scale of a wound they can be strong enough to bias cell behavior.

Many cell types can migrate directionally in electric fields, a phenomenon called *electrotaxis* or *galvanotaxis*. Keratinocytes (key skin cells), immune cells, fibroblasts, and various stem and progenitor cells have all been observed to move preferentially toward one pole or the other under physiological-strength fields in laboratory settings. Importantly, these are not exotic, damaging voltages. They can be in the same range as endogenous wound fields, making it plausible that the body uses a signal it cannot avoid anyway.

The everyday miracle of healing in your own skin likely involves such cues. When you scrape a knee, cells migrate to cover the exposed surface. Chemical attractants and clotting factors help, but the wound is also an electrical discontinuity. The cornea of the eye has been a particularly informative system for studying this, because it is transparent, accessible, and electrically active as an epithelium. Measurements in corneal wounds have placed the relevant potentials and fields in ranges that are physically meaningful for cell guidance, reinforcing the idea that endogenous electric signals are not negligible noise.

Regenerating animals appear to sustain and sculpt these signals differently.

Becker and others reported that regenerating limb stumps show a characteristic change in electrical potential after amputation: a shift that does not simply snap back to baseline, but persists. Later studies in amphibians and other model systems broadened the picture, showing that a sustained injury-associated electrical state correlates with regenerative success, while a rapid return to the pre-injury electrical condition correlates with simple closure and scarring. The phrase "short-circuiting" is not just metaphorical. If the epithelial battery is quickly repaired in a way that eliminates the distinctive wound current, the system may lose an early organizing cue that would otherwise set a regeneration program in motion.

One must be careful here. Correlation is not causation, and regeneration is not a single switch. It is an orchestra. But electrical cues have a special kind of plausibility because they are upstream of so many other processes. Voltage affects calcium entry; calcium affects gene expression and cytoskeletal dynamics; cytoskeleton affects migration and division; division and migration shape tissues; tissues then alter the electrical properties again. An injury current is therefore not merely a "signal" in the way a hormone is a signal; it is part of a feedback loop that can either stabilize a regenerative state or fall into a scar-forming attractor.

A concrete demonstration that the loop can be nudged comes from experiments in frogs. Tadpoles can regenerate tails readily, but their ability to regenerate limbs is stage-dependent and often limited. Michael Levin's group and collaborators explored whether altering the wound's ionic environment could shift the outcome. One striking approach used a wearable device sometimes described as a "bio-dome": a small cap placed over an amputated limb stump, creating a controlled microenvironment. By delivering a progesterone-containing mixture and shaping the ionic conditions at the wound, the treatment aimed to re-establish a pro-regenerative electrical state rather than allowing the wound to settle quickly into closure-and-scar mode. Reports from this line of work suggested that under certain conditions, regenerative responses could be improved, effectively coaxing a non-regenerator toward regrowth.

The details of such experiments matter: species, developmental stage, amputation level, and the exact conditions of the device all affect outcomes. Regrowing a limb is not like regrowing skin. Yet the conceptual shift is powerful. Instead of asking only which genes must be turned on, one can ask what physiological boundary conditions must be maintained so that the tissue interprets injury as a patterning problem rather than merely as damage control.

A useful image is to picture the wound as a temporary "port" into the body's control system. Through that port, ions flow. Those ion flows set voltages. Those voltages bias cell decisions. If the port is closed immediately and the electrical disturbance is quenched, the system may default to rapid sealing. If the disturbance is stabilized in a particular way, the system may recruit a more ambitious program: building, not just patching.

What would such a stabilizing signal look like over time? The exact shapes differ between experiments, but the general idea is that regenerators show a pronounced deviation in transepithelial potential after injury that persists and then settles into a plateau while regeneration proceeds, whereas non-regenerators show a deviation that collapses back toward baseline relatively quickly as the wound seals.

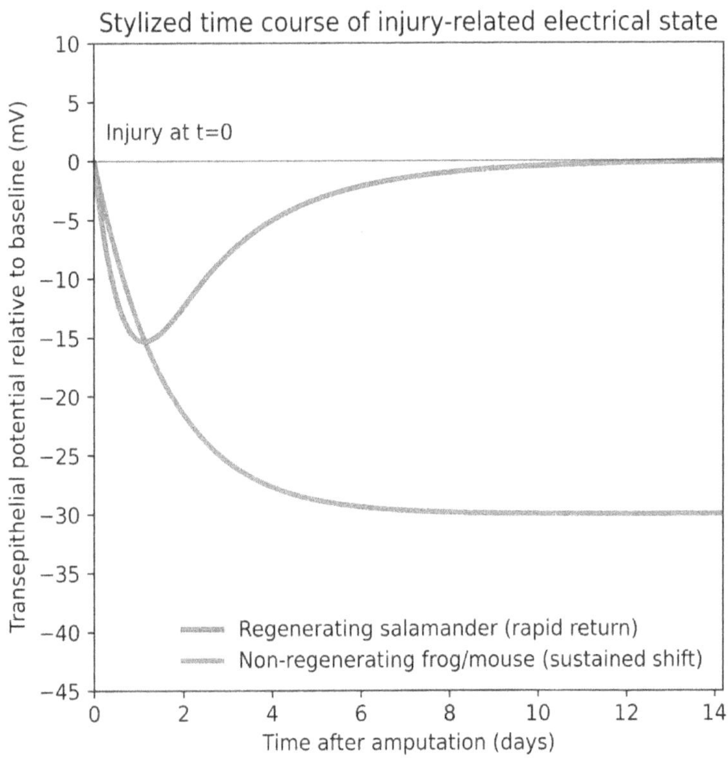

Even as a schematic, the graph suggests a different way to describe "regeneration competence." It is not only a property of genes or stem cells. It is also a property of *wound physiology*: whether the injury is allowed to become an electrically organized state long enough to coordinate downstream events.

Why would a sustained electrical state help? One reason is navigation. Cells involved in repair and regrowth must move to the wound site and orient themselves. In an electric field, many cells polarize internally: proteins and lipids redistribute to one side of the membrane, the cytoskeleton reorganizes, protrusions form preferentially at the leading edge, and migration becomes biased. This does not replace chemical cues; it interacts with them. A growth factor gradient might tell a cell that "this region is rich in signal," while the electric field tells it "move this way." In crowded, noisy tissue environments, having two orthogonal guidance systems can be the difference between a coordinated swarm and a confused milling crowd.

Another reason is identity. Regeneration is not simply about getting cells to the right place; it is about telling them what to become. Electrical states can influence gene expression patterns, in part through voltage-sensitive transport of small signaling molecules and through second messengers like calcium. A wound that sustains a particular transepithelial potential may create a permissive environment for the activation of patterning pathways that define a blastema and keep it proliferative and organized. A wound that collapses quickly back to baseline may bias the system toward differentiation and closure, resolving the emergency but foreclosing the longer project of rebuilding.

A third reason is timing. Electrical signals are fast. They appear immediately, before transcriptional programs have had time to unfold. In that sense, they can act as the opening chord that sets tempo and key. By the time genes change expression, the tissue may already have committed to a path: a regenerative trajectory or a scarring trajectory. If that is true, then therapeutic interventions that arrive late may struggle. The crucial window might be hours, not days.

There is a human emotional hook here that is hard to ignore. Anyone who has watched a loved one recover from surgery, or who has lived with a scar that tightens and aches, has felt that healing is both miraculous and incomplete. Regeneration research is often sold as fantasy medicine: regrow limbs, repair spinal cords, reverse aging. The sober reality is slower, and the ethical stakes are real. But the physiological view offers something immediately practical: even within ordinary healing, electrical cues may be harnessed to improve outcomes, reduce scarring, and guide cells where we want them.

Electric bandages and bioelectric dressings are already being explored in clinical contexts, not to regrow arms, but to help chronic wounds close and to reduce infection risk. The idea is not new. It echoes Becker's insistence that the body's own currents are part of its healing toolkit, and that medicine might do better by cooperating with those currents rather than ignoring them.

The most provocative implication, though, is conceptual rather than clinical. In a scar-forming organism, the wound is treated as a breach to be sealed as quickly as possible, restoring barrier function and extinguishing the unusual electrical state. In a regenerating organism, the wound is treated as a *problem of pattern*: a temporary, electrically active state is maintained and shaped, as if the tissue is holding the system in a special mode long enough to compute what is missing and coordinate the reconstruction.

A limb stump, then, is not merely damaged tissue. It is tissue in a different electrical regime, with different boundary conditions, and therefore different possibilities. The difference between scarring and regeneration may begin not with a gene, but with how long the body allows itself to remain electrically wounded.

4.3 Target Morphology and the Electric Scaffold

A scar is a kind of conclusion: the body has decided that "closed" is good enough. Regeneration feels like a different kind of decision. It is not satisfied by closure. It keeps asking, quietly but insistently, whether the result matches a remembered standard. A salamander does not stop at a sealed stump; it builds until the stump is no longer a stump. A planarian fragment does not stop at a blob of new tissue; it remodels until the proportions look like a proper animal. The difference is not just speed or stem cells. It is the presence of a goal.

That word makes biologists nervous. Goals sound like philosophy smuggled into physiology. Yet there is an older scientific language for this, developed not in zoology but in engineering: *cybernetics*. In the mid-twentieth century, Norbert Wiener and others tried to understand purposeful behavior without invoking spirits or intentions. A thermostat keeps a room near a set temperature; a ship's autopilot holds a course; the inner ear stabilizes gaze. None of these devices "wants" anything, but each contains a *set point* and a *feedback loop*: a way to measure the current state, compare it to a target, and apply corrections until the difference—the error—shrinks.

Regeneration begins to look less mysterious when seen through that lens. The target is not a temperature. It is anatomy: a particular arrangement of organs, symmetries, lengths, and boundaries that count as "correct" for that organism. The current state is whatever injury has left behind. The corrections are growth, cell death, migration, differentiation, and remodeling. The entire enterprise is a control problem: how to drive an enormously complex, noisy, living material back toward a stable form.

The phrase *target morphology* is a deliberately plain label for this idea. It does not claim that the organism has a picture of itself pinned to a mental bulletin board. It claims something more testable: that there exists, in the living system, information specifying what final form should be, and that the system can act to reduce the discrepancy between the current form and that specification.

Where could such a target be stored?

One answer is familiar: gradients of morphogens (signaling molecules) and networks of gene regulation. Those are real and indispensable. But they also have an awkward feature as long-term memories: they are chemically expensive to maintain and, in adult tissues, often transient. Moreover, they are usually discussed as local cues. A morphogen gradient can tell a cell whether it is "more anterior" or "more posterior," but the maintenance of a whole-body plan across time demands coordination across distances and across events. A planarian may be cut repeatedly; a limb may be amputated at different levels; tissues may be rearranged experimentally. A robust set point must survive these perturbations.

Bioelectricity offers a substrate for such robustness because it naturally forms patterns that are (i) distributed across many cells, (ii) maintained by ongoing physiological processes, and (iii) capable of settling into stable states that persist even after disturbances. In other words, tissues can behave like networks with memory.

The key ingredients are deceptively simple. Individual cells maintain membrane potentials. Cells communicate through gap junctions, forming coupled sheets and volumes. Ion channels and pumps respond to signals, changing the local voltage. Voltage, in turn, changes the behavior of channels and signaling pathways. This creates the possibility of *attractors*: stable patterns of voltage across the tissue that, once established, tend to re-establish themselves after small perturbations, much as a marble returns to the bottom of a bowl after being nudged.

Attractors are a mathematician's concept, but they have a very bodily feel when you watch regeneration. The organism behaves as if its anatomy is the bottom of the bowl: push it away by cutting, and it rolls back. The push is dramatic; the return is patient. What matters, from the cybernetic point of view, is that the return is not hard-coded as a linear chain of steps. It is a convergence toward a stable state.

A concrete example comes from the most theatrical outcome in planarians: polarity reversal and the creation of two-headed worms. When long-range communication is disrupted—for instance by interfering with gap junctional coupling, as in work by Oviedo, Morokuma, and colleagues (2009)—a posterior-facing wound can make a head. That is already striking, but the deeper significance is that the animal can be shifted into a new patterning regime that persists. Later amputations can reproduce the altered outcome, suggesting the system has adopted a different attractor: a different target morphology for what it rebuilds.

The phrase "electric scaffold" is useful here, as long as it is not taken too literally. A scaffold in construction is a physical frame. In morphogenesis, an electric scaffold is a physiological frame: a spatial pattern of voltage and connectivity that constrains where growth and differentiation are likely to occur. It is not rigid. It does not dictate cell fates by force. It biases and coordinates, providing a soft geometry in which chemical and genetic processes unfold.

The power of this idea is that it explains how an organism can be both flexible and reliable. If form were specified only by a sequence of local instructions ("if you are near the cut edge, make tissue"), it would be hard to guarantee correct global outcomes. If form were specified only by a rigid blueprint, it would be hard to accommodate unusual injuries. A distributed bioelectric pattern can do both: it can respond to context while still pulling the system toward a stable anatomical solution.

Cybernetic thinking also clarifies an otherwise puzzling feature of regeneration: stopping. Growth is easy to imagine; stopping at the right time is the marvel. Cancer, in one sense, is what happens when stopping fails. Normal development and regeneration build organs to the correct size and then cease, often leaving behind cells capable of growing again if needed but choosing not to. This is a control problem: growth must be turned off when error is near zero.

Bioelectricity intersects with stopping in a straightforward way because voltage states influence proliferation and apoptosis (programmed cell death). In planarians, bioelectric signals have been tied not only to head-versus-tail identity but also to scaling and proportional remodeling. Beane, Morokuma, and colleagues (2013) reported that manipulating ion transport can yield worms with abnormal organ proportions, implicating bioelectric state in the regulation of size and in the remodeling processes that prune and reshape tissue. Apoptosis may sound like destruction, but in morphogenesis it is often a sculptor's chisel: removing cells in the right places to restore symmetry and proportion. A control system that cannot kill cells in a disciplined way is as flawed as one that cannot grow them.

A historical anecdote helps make cybernetics feel less abstract. In the 1940s and 1950s, W. Ross Ashby, a psychiatrist and pioneer of cybernetics, built a device called the *homeostat*. It was a set of interconnected units that would adjust themselves to maintain stability despite disturbances. The homeostat was not alive, but it illustrated a principle: stability can emerge from networks that sense and correct their own deviations. Ashby was trying to understand the brain and mental illness, but the lesson travels well. A regenerating animal is a homeostat for shape. When the body is perturbed, it searches through physiological possibilities until it finds a stable configuration again.

Bioelectric networks provide the kind of adjustable connectivity Ashby admired. Gap junctions can open and close; channels can be inserted into membranes or removed; pumps can be upregulated or inhibited; neurons can contribute long-range signals that couple distant regions. The result is not a static map but a living control architecture.

The most challenging part of this story is making it feel mechanistically grounded rather than poetic. How does a voltage pattern become a growth pattern?

One route is through *electrophoresis* and *electroosmosis* at the cellular scale: electric fields can bias the movement of charged molecules and influence membrane protein distribution. Another route is through voltage-gated channels, especially calcium channels. Calcium is a universal second messenger; changes in intracellular calcium can rapidly influence gene expression, cytoskeletal dynamics, secretion, and cell-cell adhesion. Yet another route is through transporters that move not only ions but also small signaling molecules whose concentrations matter for patterning. Add to this the fact that voltage influences the state of gap junctions themselves, and one begins to see a multi-layer system: electrical patterns shape chemical patterns, chemical patterns feed back on electrical patterns, and both together shape tissue mechanics.

From the cybernetic perspective, the important thing is not which molecular lever is pulled first, but that there exists a reliable comparator mechanism: some way of turning "difference from target" into "instruction to act." In a thermostat, the comparator is a simple bimetallic strip or sensor reading. In an organism, comparators can be distributed. A tissue can detect, locally, whether it has the right neighbors, whether tension is correct, whether innervation is present, whether voltage matches a regional set point. The aggregate of those local comparisons can implement a global correction.

A reader might worry that "distributed" is a way of evading detail. It is not. Distributed control is a common engineering strategy precisely because it is robust. The internet routes packets without a central controller knowing every path. Ant colonies build nests without a single ant holding the blueprint. A tissue can coordinate patterning through local rules, as long as the rules are embedded in a global field or network state that keeps them consistent. Bioelectricity is a candidate for that global consistency because voltage patterns span tissues and can synchronize behavior over distances.

The idea of an electric scaffold becomes especially concrete when one considers experiments that "reset" patterning without changing DNA. Ion channel drugs, gap junction blockers, or genetic perturbations of ion transport can alter membrane potentials in specific regions. If the resulting animals regenerate with altered anatomy, the most economical explanation is that voltage patterns are upstream of the anatomical decision. This does not deny the role of genes; it locates genes within a hierarchy. Genes encode the components of the scaffold, but the scaffold's state can be edited by physiological inputs.

The possibility of editing target morphology raises both excitement and unease. Excitement, because it suggests a path toward regenerative medicine that does not require rewriting the genome. Unease, because it implies that the body's definition of "normal" might be more flexible than we like to admit.

Flexibility has a benign face. Consider a birth defect in which an organ is malformed. Traditional approaches aim to correct the structure surgically or to manage symptoms. A control-system approach imagines something different: restoring the correct set point so the body does more of the correction itself. If a tissue is capable of remodeling toward a target, then providing it with the right electrical boundary conditions might shift the remodeling trajectory. This is not science fiction. Orthodontics works because bone remodeling is guided by sustained mechanical forces; the body is induced to treat a new jaw configuration as the target. Bioelectric interventions might, in principle, do something analogous with certain tissues, providing sustained physiological cues that bias growth and patterning over time.

Electric fields can guide cell migration, and external fields can be applied non-invasively. That suggests a practical toolkit: electrodes, wearable devices, conductive dressings, ion-containing hydrogels, optogenetics in experimental contexts, and pharmacological modulation of channels and pumps. The goal would not be to "zap" tissue into shape, but to provide a gentle, sustained bias that cooperates with the tissue's own error-correction mechanisms.

Flexibility also has a dangerous face. If the set point can be edited, it can be edited badly. This connects bioelectricity to oncology in a way that is more than metaphor. Cancer can be framed as a failure of tissue-level governance: cells proliferate and invade in ways that ignore the organism's architectural constraints. There is evidence that bioelectric states correlate with tumorigenic behavior and that manipulating membrane potential can influence proliferation and differentiation in some contexts. Even without making strong claims, one can see why the connection is plausible. A tissue whose electrical scaffold is disrupted may lose the coherence needed for normal pattern maintenance. In a cybernetic language, the comparator becomes noisy, the error signal becomes misinterpreted, and growth runs without proper braking.

It is also worth acknowledging a more philosophical discomfort: cybernetics makes living systems sound machine-like. Yet the point is not to reduce life to machinery. It is to take seriously that living systems achieve feats—robustness, self-repair, self-limitation, and adaptive growth—that engineers struggle to replicate. Cybernetics was, in part, an attempt to learn from biology. Applying it back to regeneration is not an insult to life; it is recognition of life's competence as a control system.

A final concrete image ties the pieces together. Imagine a planarian fragment that has lost its head. At the cut edge, cells experience injury currents and altered ion flows. The local environment shifts; neoblasts proliferate; blastema tissue forms. But simultaneously, the tissue network is solving a global problem: restoring the anterior end of a body plan. If the bioelectric scaffold is in its normal attractor state, the voltage pattern near the wound shifts in a way that is consistent with "anterior," and downstream pathways follow suit. If the scaffold has been reprogrammed, the same wound can be interpreted differently, and the rebuilding converges toward a different stable solution.

This makes regeneration feel less like magic and more like a negotiation between matter and information. The matter is the stem cells and tissues. The information is not only in genes and gradients, but also in the collective electrical state of cells coupled into networks. The body heals by computing its own shape, and the computation is carried partly in voltages that are too small to feel, too ordinary to notice, and yet capable of steering flesh toward a remembered form.

Chapter 5

Rebelling Cells and Static Signals

What happens when a cell stops listening to the body's collective instructions and reverts to a primal, selfish state? This chapter examines the startling evidence that cancer is fundamentally a disorder of bioelectric communication-a loss of multicellular identity-and reveals how restoring these ancient signals can tame tumors and correct birth defects without touching the genome.

5.1 The Cell That Forgot Its Mission

A malignant tumor feels, in the lived experience of medicine, like an invasion. It crowds healthy tissue, steals nutrients, co-opts blood vessels, spreads to distant organs, and resists attempts to eliminate it. The language that grows up around cancer—*war* metaphors, "fighting" and "attacking," "magic bullets," "collateral damage"—fits the clinical drama.

Yet there is an uncomfortable biological fact behind the drama: most cancers are made of *you*. Not bacteria, not fungi, not a parasitic worm. They are your own cells, carrying much of the same molecular machinery as the cells they displace. That is precisely why cancer is so hard to treat without harming the body. A pathogen can be targeted as foreign; a renegade cell must be addressed without burning down the whole house.

One way to sharpen this distinction is to ask a question that sounds philosophical but turns out to be practical: what makes a cell a good citizen of a body? Genes matter, of course. So do chemical growth factors, mechanical forces, immune surveillance, oxygen, inflammation. But a multicellular body also depends on something less often granted center stage: the fast, quiet, continuous *electrical conversation* that cells conduct across their membranes and through their connections to neighbors.

Every living cell is an electrical object. Its membrane is a thin insulating film separating two salty solutions—inside and outside—each containing charged particles (ions). Because the membrane is selectively permeable and because pumps and channels move ions unevenly, the inside of the cell typically sits at a different electrical potential than the outside. This difference is the *membrane potential*, often written V_{mem}. If you imagine a tiny voltmeter with one lead inside the cell and one outside, V_{mem} is what it reads.

Neurons made bioelectricity famous because they use rapid spikes of voltage to transmit signals. But most cells are not trying to think in milliseconds. Their electrical life is slower, more like a sustained posture than a sudden shout. Even so, that posture—how polarized or depolarized a cell is—can influence whether it divides, migrates, differentiates into a specialized type, or participates in building and maintaining tissue architecture.

A useful shorthand is this: when the inside of a cell is more negative than the outside, the cell is *polarized* (often, *hyperpolarized* if it is especially negative). When that negativity is reduced—when the voltage drifts toward zero—the cell is *depolarized.* These are not moral categories. They are electrical states. But biology has learned to use them as signals.

Cancer biology has long been dominated by genetics: mutations in oncogenes, loss of tumor suppressor genes, genomic instability, chromosomal rearrangements. That framework has produced enormous insight and real therapies. And yet, even in a world of genomics, it remains striking that tissues can keep order despite the fact that mutations arise constantly in ordinary life. A liver cell, a skin cell, an intestinal cell—each divides, repairs, and endures chemical insults for decades. If a single mutation were sufficient to unleash malignancy, bodies would be unrecognizable within a lifetime.

Order is maintained not only by policing individual molecules but by enforcing collective rules. A cell in a body is not merely a bag of enzymes carrying a genome. It is a participant in a coordinated project: maintaining shape, respecting boundaries, sharing resources, limiting proliferation, dying when necessary, specializing when necessary, and repairing damage without improvising new anatomy. Cancer can be viewed as a collapse of that multicellular agreement—a cell that stops taking calls, stops listening, and begins to behave as if it were alone in the world.

There is a concrete cellular structure that turns this metaphor into something you can point to under a microscope: the *gap junction*. Gap junctions are protein-built channels that directly connect the interiors of adjacent cells. Each channel is formed by connexin proteins (in vertebrates) arranged into a pore that spans both membranes, creating a tiny bridge. Through these bridges, small molecules and ions can pass: not everything, not large proteins, but enough to share electrical and chemical states. A tissue laced together with open gap junctions is, in a literal sense, a network. When many gap junctions close, the network fragments into isolated compartments.

The physiological consequence is easy to picture. Suppose a neighborhood of cells needs to make a decision—how fast to grow, which direction to migrate, when to stop. In a well-coupled tissue, signals can spread, averaging local noise and enforcing a shared plan. When coupling is lost, each cell becomes more like a solitary organism. It must solve problems alone: find nutrients, resist death, exploit opportunities to divide. Many of the behaviors that horrify us in cancer—unbounded proliferation, refusal to die, invasion—make grim sense as survival strategies for a cell that experiences itself as cut off from the body's governance.

This is not merely a metaphorical reframing. Michael Levin's group and collaborators, working especially in frog embryos, has argued that malignant transformation can have an electrical signature: cells that will participate in tumor-like growth often show *depolarization* before the tumor is visible. The electrical change arrives early, like a warning light on a dashboard. It suggests that cancer can be, at least in part, a disorder of bioelectric patterning—of the voltages and couplings by which tissues maintain their identity and geometry.

A historical echo makes this idea feel less alien than it first appears. In the early twentieth century, long before anyone could sequence a genome, researchers argued over whether cancer was primarily a *local* phenomenon—a derangement inside a cell—or a *tissue* phenomenon—a failure of the surrounding environment to keep order. The question was pursued with the tools of the time: transplantation experiments, tissue culture, microscopy, and physiology. The genetic revolution tilted the balance toward cell-intrinsic explanations. But the earlier tissue-centric intuition never entirely disappeared, because clinicians kept seeing puzzles that genetics alone did not neatly answer: tumors with similar mutations behaving very differently, dormant micrometastases that wake up years later, and cases where changing the tissue context seemed to change the fate of abnormal cells.

Bioelectricity offers a way to take tissue-level control seriously without mysticism. Voltage is a physical variable; coupling is a physical variable. Both can be measured, perturbed, and mapped.

Consider what it means, physically, for a cell to be depolarized. Cells maintain their membrane potential with ion pumps (like the sodium-potassium pump) and with ion channels that allow specific ions to leak or flow under certain conditions. Potassium, sodium, chloride, hydrogen ions—all have their own gradients and pathways. Changing the expression of channels, altering their gating, or perturbing pumps can shift V_{mem}. A depolarized cell is not simply "missing electrons." It is in a different electrochemical regime, and that regime influences other systems. Voltage affects the movement of charged molecules, the opening of voltage-gated calcium channels, the distribution of signaling lipids, and the activity of transporters that import or export metabolites. It can also influence gene expression indirectly, by changing calcium dynamics and downstream transcription factors.

From the cell's point of view, V_{mem} is part of its internal weather: it shapes what it can do next.

Now add gap junctions. A single cell's V_{mem} is important, but a tissue's *pattern* of voltages—spatial gradients, domains of relative hyperpolarization and depolarization—is potentially even more important. Pattern is where information lives. A uniform field says little; a structured field can instruct.

When a tissue is electrically coupled, a voltage change in one region can influence distant regions, smoothing out abrupt discontinuities or propagating coordinated shifts. When coupling is disrupted, local voltage states become private, and private states can drift.

This is the intuition behind describing cancer as a cell that "forgot its mission." The "mission" is not a moral calling; it is the role assigned by the body's patterning networks. In a healthy adult tissue, those networks maintain differentiation (cells keep doing what they are meant to do) and enforce appropriate proliferation (cells divide when needed, stop when not). In development, those networks build anatomy from scratch. In both cases, bioelectric signals are part of the control layer that sits between raw molecular machinery and the large-scale geometry of organs.

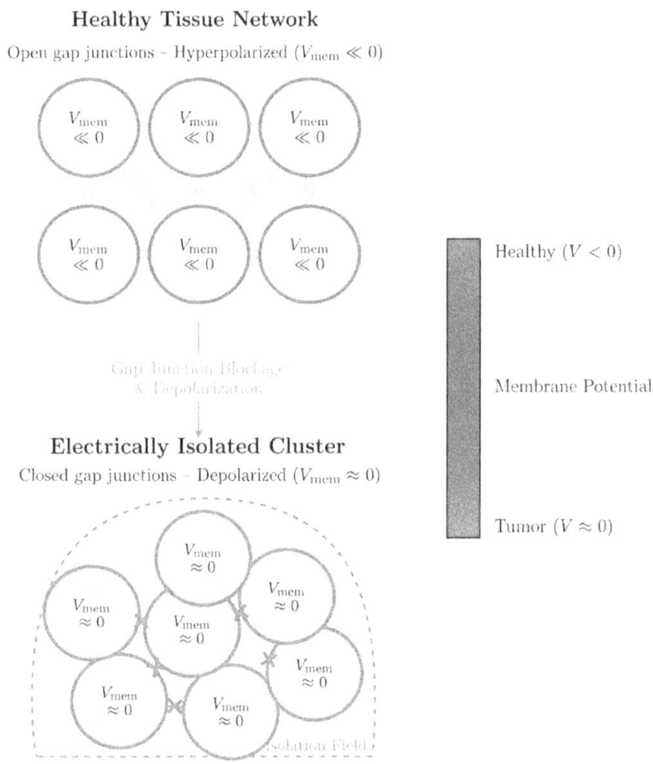

The diagram is deliberately simplified, but it captures three ideas that are easy to miss when cancer is described only in terms of mutations. First, tissues are *networks* before they are masses: their cells exchange not just hormones and cytokines through extracellular space, but also direct currents and small messengers through cell-to-cell bridges. Second, malignant clusters can be *crowded but isolated*. They touch neighbors, press against them, even exchange soluble factors with them, yet become electrically and metabolically less coupled. Third, depolarization is not just an internal state; it can be a flag that a cell is slipping out of the collective regime.

A concrete example makes these abstractions feel less like wordplay. In one set of experiments associated with Levin's lab (2010), a small population of "instructor" cells in frog embryos was electrically altered in a way that did not directly mutate the pigment cells that later misbehaved. The surprising outcome was that melanocytes—pigment cells—elsewhere adopted neoplastic-like behaviors: excessive proliferation, abnormal migration, and invasive colonization. The relay, as worked out in that work, involved serotonergic signaling: changing voltage altered the handling of serotonin, which then influenced distant cells. The point is not that serotonin is "the cancer molecule," or that frogs are humans in miniature. The point is that voltage state can be upstream of biochemical cascades and can act at a distance, and that a tissue can contain electrically defined "decision-making" domains that steer cell behavior non-cell-autonomously.

If that sounds unsettling, it should. It implies that a cell's fate is not always sealed by its own genome. It can be pushed toward normality or abnormality by the electrical state of its environment, including cells it never touches directly. Cancer, in that view, is not merely a story of bad DNA; it is also a story of a failed conversation in a multicellular society.

The phrase "failed conversation" is not meant to be gentle about the disease. Tumors kill, and for the person facing a diagnosis, conceptual elegance offers little comfort. But reframing can change where one looks for leverage. If a tumor is an enemy army, the only sensible response is to destroy it. If a tumor is a population of your own cells in a pathological state, then the therapeutic imagination broadens: one might kill, yes—but one might also *coerce*, *re-educate*, or *reconnect*.

To understand what "reconnect" could mean, it helps to be precise about what gap junctions do and do not do. They are not simply holes that make all cells one big bag of cytoplasm. They are regulated channels; they open and close in response to pH, calcium, phosphorylation, and other signals. They pass some molecules but not others. They can couple electrical potentials so that groups of cells share a common voltage landscape. That coupling can enforce stable tissue states in the way that a flock maintains formation: not by central command, but by local interactions that create coherent global behavior.

Cancer cells often show altered connexin expression or localization, and reduced functional coupling. The relationship is complex—connexins can suppress tumor formation in some contexts yet aid later steps like metastasis in others, because biology reuses tools for multiple purposes. But the basic observation that many tumors are less well-coupled to surrounding tissue is robust enough to sustain a provocative hypothesis: losing gap junctional "membership" may be part of how a cell stops acting like an organ component and starts acting like a free-living organism.

Electrical disconnection is rarely absolute. Tumors still exchange signals with their surroundings; in fact, the tumor microenvironment is famously interactive. Cancer recruits blood vessels, shapes immune infiltration, remodels extracellular matrix, and alters metabolism around it. The claim here is narrower and more testable: that the *bioelectric state*—the resting voltage and the coupling that allows voltage patterns to exist across tissue fields—can be an upstream control variable, and that shifts in that variable can predispose tissue toward tumorigenesis.

A clinician might ask, reasonably, why this matters if mutations still drive malignancy. The answer is that biology often has multiple layers of control, and layers can override one another. An embryo can build a correct body plan even when individual cells vary in gene expression; tissues compensate. Likewise, a tissue might be able to suppress the malignant potential of mutated cells if the collective pattern-control signals are strong and coherent, while the same mutations might blossom into a tumor in a tissue whose electrical governance has been weakened by chronic inflammation, injury, aging, or other stressors that perturb coupling and membrane potential.

This way of thinking also makes emotional sense of something patients often feel but rarely say without being misunderstood: that cancer can feel like a betrayal from within. It is not that the body "turned against itself" in a moral sense; it is that a few cells slipped out of the body's shared rules. They stopped respecting the quiet constraints that normally make tissues livable places.

If bioelectricity is part of those constraints, then "good citizenship" has a measurable physiology. A well-differentiated tissue tends to maintain stable voltage domains and robust coupling that allow positional information and growth control to spread. A rebel cell tends to drift toward depolarization and disconnection, becoming more self-referential, less governed by the tissue's pattern memory.

The deepest value of this picture is not that it replaces genetics, but that it restores a missing scale of explanation: the scale at which a body is a coordinated electrical-chemical structure, not merely a collection of cells each running its own program. Cancer, seen from that scale, is not an invader from outside. It is a breakdown of belonging—an electrical quieting of the signals that tell a cell who it is, where it is, and what it owes its neighbors.

5.2 Coercing the Rebel

There is a particular kind of dread that comes with hearing the word *metastasis*. A tumor in one place is frightening; a tumor that has learned to travel feels like a story that has left the page. Modern oncology has built powerful weapons for that fear: surgery, radiation, chemotherapy, targeted drugs, immunotherapy. Many of these approaches share a basic premise: the malignant cell is dangerous, so the goal is to remove it or kill it.

That premise is often right. The trouble is that it is not always the only premise available.

A different, quieter possibility has been hovering at the edges of cancer biology for over a century: that some malignant behavior is not a permanent identity but a state-one that can be pushed, pulled, and sometimes reversed by changing the cell's surroundings. The idea has an old, almost quaint name: *tumor normalization*. It sounds naive until one remembers a basic fact of life: cells are not just bags of genes. They are responsive systems embedded in tissues, and tissues are governed by signals that operate at scales larger than any single cell.

The embryo provides a useful starting image because it is the most dramatic example of collective governance. An embryo can take a fertilized egg and, without a supervisor, produce a body in which the heart forms in the chest and the eyes form on the face. That success depends on rules that are not written in one place; they emerge from networks of communication—chemical, mechanical, and electrical. Adult tissues retain a more modest version of the same capacity: they maintain architecture, enforce boundaries, and repair damage. Cancer can be read as what happens when those rules lose their grip on a subset of cells.

The word *microenvironment* is often used to describe the neighborhood that surrounds tumor cells: nearby stromal cells, immune cells, extracellular matrix, oxygen levels, and soluble factors. The microenvironment can nourish tumors, but it can also restrain them. If malignant cells behave partly like rebels against tissue-level governance, then changing the microenvironment is not merely supportive care. It becomes an attempt to make the rebel legible again—to restore the signals that say, in effect, "you are part of this body."

Bioelectricity offers a specific version of that strategy: rather than aiming only at mutations inside tumor cells, one can try to reset the *electrical setpoints* of the tissue field that surrounds them. The key concept is deceptively simple. Every cell maintains a resting membrane potential, V_{mem}, determined by the activity of ion channels and pumps. Many cells are connected by gap junctions—direct bridges that allow ions and small molecules to pass between cells, coupling them into a shared electrical domain. A tumor is not only a clump of proliferating cells; it is a distortion in a broader electrical landscape.

If that landscape influences whether an oncogene-driven cell will actually form a tumor, then the neighborhood becomes part of the therapy.

This sounds like an attempt to replace molecular precision with mystical "energy fields." It is nothing of the sort. Tissue bioelectricity is a measurable physiology, and its manipulation can be blunt or precise. Ion channel drugs are already routine in medicine: antiarrhythmics target cardiac channels; anticonvulsants target neuronal excitability; diuretics target renal transporters. The novelty here is not that ion channels matter, but that their state can encode instructive information about tissue patterning and that shifting voltage domains can change the trajectory of oncogene-expressing cells.

A concrete example comes from frog embryos, which have become a kind of proving ground for this line of research because their development is accessible, rapid, and tolerant of experimental intervention. Brook Chernet and Michael Levin (2014) reported that oncogene-mediated tumor-like structures in *Xenopus* are influenced by the endogenous membrane potential of host tissues, even at long range. In other words, a cell's genetic push toward tumorigenesis was not the whole story; the bioelectric condition of the surrounding embryo altered the likelihood and character of tumor formation.

That "long-range" detail is the part that forces a reader to stop and re-evaluate. It is easy to accept that local chemical signals influence a tumor. It is harder—but arguably more revealing—to accept that distant tissue states can set permissive or restrictive conditions. Long-range influence implies that a body maintains something like a distributed governance system: not a single signal, but a coupled network that can broadcast constraints.

Gap junctions are one plausible conduit for that broadcast. When cells are electrically coupled, local changes in ion flux can influence distant regions, and small messenger molecules can travel through the junctions. Chernet and Levin (2015) went further by showing that disrupting gap junctional communication could reduce oncogene-mediated tumor formation in *Xenopus* embryos, including when the disruption was not limited to the tumor region itself. The implication is not that gap junctions are always "good" or always "bad"—cancer biologists have long noted that connexins can suppress early tumor growth yet facilitate later steps such as metastasis in some contexts—but that coupling is a control knob. Turn it, and the system's behavior changes.

If a tumor is a breakdown of multicellular coordination, then it makes sense that altering the coordination channels can change the fate of the would-be tumor.

A historical anecdote helps anchor this in a broader tradition of "context matters." In the 1970s, the developmental biologist Beatrice Mintz performed experiments in which malignant teratocarcinoma cells were introduced into early mouse embryos. In some cases, those cancer cells contributed to normal tissues in the resulting animals, as if the embryonic environment had reined them in. The experiments were technically challenging and their interpretation is nuanced—teratocarcinoma cells have peculiar properties—but the cultural impact was clear: malignancy was not necessarily a one-way street. Under certain conditions, an abnormal cell could be coaxed back into a cooperative developmental program.

Bioelectric normalization is a modern, mechanistic cousin of that intuition. The embryo is an especially strong governor because its patterning cues are loud, coherent, and globally integrated. The adult body's cues can be quieter and more fragmented, which may be why cancers become more common with age and chronic tissue stress. But the principle remains: when the governance network is strong, even dangerous cells can be constrained; when it weakens, cells with the wrong internal incentives can take advantage.

The therapeutic promise is tantalizing precisely because it shifts the emotional and practical stakes. Killing cancer cells is hard because they are similar to healthy cells. Coercing them—pushing them into differentiation, restoring growth control, making them rejoin tissue-level constraints—suggests treatments that might be less toxic and less evolutionarily fragile. A therapy that relies on killing creates an arms race: any cell that survives becomes the seed for resistance. A therapy that relies on normalization changes the fitness landscape. The malignant strategy of rapid proliferation may no longer be advantageous if the tissue environment enforces differentiation and stable membership.

This idea does not require a denial of genetics. Oncogenes still matter.

Mutations still accumulate. A cell with a KRAS mutation does not cease to have that mutation because its voltage changes. The question is different: does the mutation *express itself as cancer* when the cell is embedded in a strongly governed bioelectric tissue field? Or can that field force the cell to behave, at least for a time, like a cooperative citizen?

One way to make that question experimentally tractable is to induce tumor-like structures using known oncogenes and then manipulate the host tissue's bioelectric state with ion-channel drugs or other tools. If tumors diminish or fail to form despite the oncogenic drive, it suggests that tissue-level physiological state is not merely a consequence of cancer but can be an upstream cause, or at least an upstream constraint.

The phrase "channel drugs" can be misleading because it conjures a picture of a single magic molecule binding a single channel. In reality, bioelectric states are often *degenerate*: multiple different channels and pumps can produce the same V_{mem}. That is frustrating for pharmacology if one expects a single target, but it is encouraging for physiology because it implies robustness. If what matters is the voltage state, there may be many ways to reach it. The body itself uses that flexibility; it does not depend on a single protein for something as fundamental as cell polarity. For therapy, it opens the possibility of repurposing existing drugs or combining modest interventions to shift a tissue's electrical regime.

A reader might reasonably ask what "shifting a tissue's electrical regime" could mean in practice. One can start with an intuitive analogy that stays grounded in biology: consider cardiac arrhythmia. The heart is an electrical syncytium of coupled cells. When coupling or excitability is disrupted, waves of excitation become chaotic. Treatments include drugs that alter ion channel conductance and procedures that alter tissue structure. The goal is not to "fix a gene" in each cardiomyocyte; it is to restore a stable electrical pattern at the tissue level. Cancer is not an arrhythmia, but the logic of tissue-level electrical stabilization is not alien to medicine.

Optogenetics provides another bridge between experimental biology and clinical imagination. Optogenetics—using light-activated ion channels or pumps to control cell voltage—was developed largely for neuroscience, where precise control is essential. But it also offers a way to ask clean causal questions in non-neural tissues: if one forces a patch of cells to hyperpolarize or depolarize on demand, what happens to growth, differentiation, and patterning?

Levin's group and collaborators have used such approaches in regeneration research: for example, optogenetic hyperpolarization via archaerhodopsin, a light-driven proton pump, has been used to rescue patterning defects caused by perturbing endogenous proton flux and to restore regenerative capacity within a defined post-injury window, with controls suggesting that voltage itself—not merely pH—was a key causal variable. The relevance to cancer is conceptual rather than direct: it shows that voltage manipulation can act as an instruction, not merely a byproduct. It shows that tissue state can have "critical windows" in which an intervention has outsized influence. It shows that bioelectricity can sit upstream of complex anatomical outcomes.

Cancer biologists are accustomed to the idea of a critical window: early lesions are often more malleable than advanced tumors, and microenvironmental constraints can suppress or permit progression. Bioelectricity provides an additional dimension to that window: a lesion may become less governable as it closes gap junctions, remodels its extracellular matrix, and establishes a depolarized enclave. If so, then a bioelectric intervention aimed at restoring coupling or normalizing voltage might be most effective early, before the lesion has insulated itself.

This is where the social metaphor of "coercing the rebel" becomes more than rhetoric. Coercion, in this context, does not mean brutality. It means restoring constraints that make cooperation the easiest path. In developmental biology, differentiation is often enforced not by threats but by context. A cell in the right neighborhood receives signals that stabilize a particular identity. Remove it from that neighborhood and it can drift. Put it back, and it may settle again.

The tumor microenvironment can be thought of as a neighborhood that has lost its ability to enforce stable identities. Hypoxia, inflammation, disrupted extracellular matrix, abnormal mechanical stiffness, and altered metabolism all contribute. Bioelectric dysregulation may be both a cause and a consequence: depolarization can promote proliferation and migration; proliferating, stressed cells can alter ion flux and coupling. The causal arrows are likely to form loops. But loops are not an obstacle to intervention. They are an opportunity. In a looped system, a small push at the right node can shift the whole system into a different stable state.

A skeptic might object that this is too optimistic: cancers are heterogeneous, evolving populations, and "normalizing voltage" sounds vague compared to sequencing a tumor and targeting its driver mutations. The best answer is that both perspectives can be true. Genetic drivers are real, and targeted therapy can be life-saving. But targeted therapy also reveals the limits of a purely cell-intrinsic view: resistance evolves quickly, and tumors often find alternative pathways. A therapy that alters tissue-level governance may be harder for a single cell to evade, because it changes the rules of the game rather than blocking one move.

There is also a diagnostic implication hiding in the therapeutic one. If malignancy correlates with bioelectric signatures—particular patterns of depolarization, altered coupling, dynamic voltage fluctuations—then one can imagine stratifying tumors by electrical phenotype as well as genetic profile. Work on voltage imaging in human cancer cell lines, including breast cancer (2022), suggests that malignant cells can have distinctive membrane potential dynamics tied to specific conductances. Even if such measurements never become routine in clinics, they sharpen the idea that cancer cells live in a different physiological state, not merely a different genetic state.

That difference matters because physiology can sometimes be changed more rapidly than genes can. A drug can shift ion flux in minutes. A change in coupling can alter collective behavior quickly. A tissue-level electrical intervention might not need to be permanent to be effective; it might need only to push a system across a threshold, after which other stabilizing processes—differentiation programs, immune surveillance, contact inhibition—can take over.

It is worth being honest about what this approach is *not*. It is not a promise that one can treat advanced metastatic cancer with a simple "voltage pill." It is not an argument that mutations do not matter. It is not an invitation to replace evidence-based oncology with vague talk of "energy." The value of bioelectric normalization is that it generates testable, mechanistic hypotheses about how the microenvironment constrains malignancy and how one might restore those constraints using tools that biology already understands: channels, pumps, and junctions.

A practical vision begins to emerge. Imagine a future in which an early lesion is evaluated not only for its mutations but for the bioelectric condition of its surrounding tissue field: is the neighborhood strongly coupled or fragmented? Is the lesion depolarized relative to its neighbors? Are there electrical gradients that suggest a loss of pattern memory? A clinician might then choose a combination therapy: a targeted drug for a known driver mutation, an immunotherapy to enhance surveillance, and a physiological intervention—perhaps an ion channel modulator or a connexin-targeting strategy—aimed at restoring the tissue's ability to enforce differentiation and growth limits.

The human appeal of that vision is not merely that it could be gentler. It is that it makes cancer feel less like an alien enemy and more like a failure of internal governance that can, at least sometimes, be repaired. The body is not helpless in the face of cellular misbehavior; it has systems for coordination. If those systems can be strengthened—if the neighborhood can be made orderly again—then even a genetically compromised cell may find itself living under rules it can no longer ignore, dividing when permitted, specializing when required, and relinquishing the lonely freedoms that define the rebel.

5.3 Static in the Developmental Channel

A newborn's face is a triumph of coordination. Bones meet at precise angles, the palate seals, the eyes sit symmetrically in their sockets, nerves find their targets, vessels thread through tissue without tangling. All of this is built while the embryo is still small enough to rest on a fingertip. When that coordination fails, even slightly, the result can be devastating: a cleft lip that complicates feeding and speech, a heart defect that demands surgery, a spine that does not close, organs mirrored to the wrong side of the body.

It is tempting to imagine birth defects as mechanical mishaps-pieces assembled in the wrong order-or as the straightforward output of flawed genes. Sometimes they are. But the more one learns about development, the more one confronts an unsettling truth: the genome is not a blueprint in the architectural sense. It is more like a library of parts, sensors, and rules for how cells respond. The actual construction depends on communication.

Chemical signals, like morphogens and growth factors, are the most familiar carriers of that communication. Mechanical forces-tension, stiffness, pressure-are now recognized as equally fundamental. Bioelectricity adds a third layer: a pattern of voltages and ion flows that can precede visible anatomy and help instruct it. Like any channel that carries information, it can also carry noise. When bioelectric signals become noisy-blurred, scrambled, or mistimed-cells may still execute their local programs, yet the collective outcome can be anatomically wrong.

"Noise" here does not mean random sparks or external electricity. It means internal physiological variability that should be structured but is not. A radio tuned slightly off-station produces static, not silence: bits of music arrive, but the message is corrupted. Developmental bioelectricity can fail in the same way. A tissue may still generate voltages, still open channels, still pump ions, but the spatial pattern and timing that normally guide construction may be degraded. The embryo then builds with a distorted reference signal.

The earliest moments of an embryo's life are not quiet. Cells divide rapidly; membranes are remodeled; ion channels and pumps redistribute; gap junctions open and close. At first glance it seems implausible that such turbulence could support reliable information. Yet reliability is what embryos achieve. One clue is that the embryo does not wait for complex gene expression patterns to appear before it starts making decisions. It uses faster signals to break symmetry, establish axes, and demarcate territories. Ion flux is fast, and membrane potentials are immediate. A cell can change its electrical state in seconds; gene expression changes typically take longer. Bioelectric prepatterns-spatial domains of different resting membrane potentials-can therefore act as early scaffolds, setting up conditions that later biochemical and genetic machinery elaborates.

Because of that timing, bioelectric disruption can be uniquely treacherous. A teratogen-a developmental toxin-does not need to mutate DNA to cause a defect. It can perturb ion transport at a critical moment, and the embryo may never recover its original map. The defect that appears days or weeks later can then look like a "genetic" problem because it is stable and patterned, yet the initial trigger was physiological.

This helps explain why some birth defects are maddeningly inconsistent. Two embryos with the same mutation can show different severities; identical twins can diverge; a population exposed to the same chemical can show a spectrum of outcomes. Development is a dynamical system: it has buffers and redundancies, but it also has thresholds and tipping points. Bioelectric patterning, sitting upstream of many other processes, can be one of the places where small perturbations get amplified.

A particularly vivid set of examples comes from *Xenopus* embryos, in which researchers can image voltage patterns across the body surface using fluorescent voltage reporters. These reporters act like tiny lanterns whose brightness or color depends on membrane potential. With them, one can watch voltage domains form and move across the developing embryo, revealing an otherwise invisible physiology. Michael Levin's group has emphasized a key implication of this visibility: bioelectric signals cannot reliably be inferred from fixed tissue samples or from gene expression profiles alone, because channel activity is controlled post-translationally and dynamically. A cell may express an ion channel gene, yet the channel can be closed, phosphorylated, or otherwise gated. The information is in the *state*, not merely in the parts list.

That distinction matters for medicine. It suggests that some developmental disorders might be "invisible" to approaches that focus only on DNA sequence and static molecular snapshots. The relevant defect may be a mis-set electrical state during a short window early in development.

One of the most compelling cases of bioelectric control involves craniofacial development. Faces are not sculpted by a single gene or signal; they are assembled by coordinated growth and folding of multiple tissues. In *Xenopus*, patterned activity of the H^+-V-ATPase-a proton pump best known from cell biology textbooks for acidifying compartments-helps generate regional differences in voltage and pH across the ectoderm. In work associated with Levin's group (2011), disrupting this pump disrupted both the electrical regionalization and craniofacial morphogenesis, along with downstream gene expression domains. In other words, when the pump was perturbed, the face did not merely grow more slowly; it grew *differently*. The embryo had lost an organizing signal.

This is a useful example because it breaks a common intuition: that ion pumps are housekeeping devices and developmental patterning is a separate, higher-level process. The embryo does not respect that division. It uses whatever physical variables are available to represent information. A voltage gradient is a variable. A pH domain is a variable. If those variables are stable and spatially organized, they can serve as coordinates for later events.

A second, almost cinematic example comes from left—right patterning, the process by which organs end up on consistent sides: heart on the left, liver mostly on the right, and so on. Left—right asymmetry is not a simple matter of flipping a genetic switch. The embryo begins essentially symmetrical. It must break that symmetry reliably and transmit the resulting "leftness" and "rightness" across developing tissues. Classical accounts emphasize cilia-driven flow and asymmetric gene expression. Bioelectric accounts add an even earlier layer: asymmetric localization and activity of ion transporters, including proton pumps, can create voltage and ion-flux differences that bias downstream signaling.

The striking feature of such early bioelectric asymmetry is that it can be selectively scrambled. Experiments manipulating H^+ flux early in development have been reported to randomize organ situs-some embryos develop normal organ placement, others show mirror-image placement or mixed arrangements-without causing a generalized collapse of development. That selectivity is a hallmark of a communication channel: you can introduce noise into one message (left—right orientation) while leaving other messages intact. It is hard to make sense of this if one imagines development as a single monolithic program. It makes much more sense if one imagines multiple information streams, each with its own vulnerabilities.

This notion of "static" also makes sense of a clinical observation that has long puzzled families and physicians: the same teratogen can produce defects in different organ systems depending on timing. Thalidomide, the infamous example from the late 1950s and early 1960s, caused limb defects when taken during specific windows of pregnancy. Timing is everything because different patterning messages are broadcast at different times. If bioelectric cues are among the earliest broadcasts, then early exposures that perturb ion transport can have consequences that appear later as structural defects.

The most hopeful part of the bioelectric story is not merely that it provides another cause of disorders. It offers a different style of remedy. If a defect arises because the embryo received a corrupted instruction signal, one can ask whether the correct signal can be re-imposed. This is not science fiction. Development already contains examples of correction: embryos can regulate, adjusting to perturbations, compensating for lost cells, reshaping despite damage. Bioelectric interventions aim to assist that natural regulation by restoring the missing coordinate system.

Eye development provides a clear illustration because it is both visually striking and mechanistically tractable. In a widely discussed set of experiments, Pai and colleagues (2012), working in the context of Levin's lab, showed that altering membrane potential could influence eye field patterning in *Xenopus*. By manipulating ion channels and thus voltage states, they could induce ectopic eyes-well-formed eyes in locations outside the normal eye field. The lesson is not that voltage is a magical "eye switch." It is that voltage can act upstream of gene regulatory networks that define organ identity. If a voltage pattern can instruct eye formation in a new location, then voltage patterns are carrying positional information in a way that is at least partly independent of the embryo's usual chemical gradients.

The bridge from voltage to gene expression is not mysterious. One well-characterized route is calcium. Many channels are voltage-gated, including channels that allow Ca^{2+} to enter the cell. Calcium is a powerful second messenger; cells use transient calcium signals to regulate transcription factors, cytoskeletal dynamics, secretion, and more. A shift in V_{mem} can therefore alter calcium entry, which alters downstream gene expression. In that sense, bioelectricity does not compete with molecular biology; it feeds into it. It changes the state of the cell in ways that molecular machinery can read.

Once one accepts that voltage patterns can instruct organogenesis, "rescue" becomes a reasonable concept. If a teratogen or mutation pushes the voltage pattern into the wrong regime, perhaps a drug or an optogenetic tool can push it back. The potential advantages are striking. Genetic mutations are difficult to reverse in an embryo; voltage states can be shifted quickly. A physiological intervention does not need to edit every cell's DNA. It needs to restore a pattern that cells can collectively interpret.

A powerful demonstration of this logic comes from regeneration research, where bioelectric interventions have been used to restore developmental capacities after injury. Optogenetic hyperpolarization using archaerhodopsin, a light-driven proton pump, has been used in experiments to rescue patterning defects caused by perturbing endogenous proton flux and to restore regenerative capacity when applied in a specific early time window after injury. Control experiments have been designed to argue that voltage itself-not merely pH changes associated with proton pumping-was the causal driver. For a reader trying to understand birth defects, the relevance is direct: embryos, like regenerating tissues, can be steered by bioelectric state during windows when patterning decisions are being made.

The phrase "override genetic errors" requires careful handling because it can be misunderstood as a promise of miraculous correction of any mutation. Biology rarely offers such absolutes. A lethal mutation that prevents a vital protein from ever being made cannot be erased by voltage. But many developmental outcomes are not pinned to a single gene; they emerge from networks with redundancy. In such networks, a mutation may push the system toward a different attractor state-a different stable outcome-without making the correct outcome physically impossible. Bioelectric interventions, by shifting upstream physiological conditions, may sometimes push the system back toward the desired attractor, allowing the embryo's own regulatory machinery to complete the construction.

This is not entirely alien to clinical medicine. Consider phenylketonuria (PKU), a genetic disorder that can cause severe intellectual disability if untreated. The gene defect remains throughout life, yet dietary management can prevent the devastating outcome. The broader lesson is that genotype does not dictate phenotype in a rigid way; physiological context matters. Bioelectric rescue proposes a developmental analogue: the genetic predisposition may remain, yet the anatomical outcome can sometimes be improved by manipulating early physiological state.

The challenge, of course, is specificity. Electricity is easy to imagine as global-apply a current and everything is affected. Endogenous bioelectricity is not like that. It is patterned, cellular, and local. The embryo's voltages are produced by particular pumps and channels in particular places. A useful therapeutic intervention would need to be similarly patterned, or at least timed and targeted in a way that nudges the correct tissues without disrupting others.

Timing may be the more accessible control lever. Development is full of critical windows. If bioelectric patterning operates early, then interventions could focus on those early windows, perhaps even transiently, to set the embryo onto a stable developmental path. The prospect of prenatal intervention raises ethical and practical issues that cannot be ignored, but even without clinical translation, the experimental finding itself changes how one thinks about causality. A birth defect is not always a static consequence of damage; it can be the endpoint of an information-processing failure, and information-processing failures can sometimes be corrected by restoring the signal.

There is also a sobering corollary. If bioelectric signals guide development, then environmental factors that subtly affect ion channels and pumps may have underappreciated developmental consequences. Many common drugs act on ion channels, sometimes unintentionally. Nutritional states influence electrolytes. Inflammation alters tissue pH and ion transporter activity. The embryo is buffered, but not invulnerable. A physiology-based view of development expands the list of relevant exposures beyond the classic DNA-damaging agents.

For all the sophistication of modern genetics, one of the persistent mysteries of embryology is reliability: how embryos achieve near-perfect anatomy so often, despite noise in molecular processes. One answer is that they use multiple layers of control, each capable of smoothing noise in the others. Bioelectricity may be one of those layers, a distributed system of state variables that cells can read quickly and collectively. When that system is clean, development proceeds with astonishing grace. When it becomes noisy, the embryo may still build, but it may build the wrong thing with great confidence.

A final image returns to the human scale. Parents of a child with a congenital disorder often ask a question that has no easy emotional landing: "Why did this happen?" Sometimes the answer is a known mutation. Sometimes it is an exposure. Often it is neither. The bioelectric perspective does not supply comforting certainty, but it does supply a different kind of intelligibility. It suggests that some disorders may begin as fleeting physiological disruptions-moments when the embryo's internal signals were garbled-and that those moments can cascade into stable anatomical differences. It also suggests, with cautious optimism, that because the disruption is physiological, it may be reachable by physiological means: not by rewriting the genome, but by restoring the signals that tell cells how to cooperate in building a body.

Chapter 6

Hacking the Operating System

Until recently, the electric conversations of cells were invisible, hidden behind the limitations of single-electrode probes. Today, we stand on the brink of a revolution where light can both illuminate and rewrite the voltage gradients that shape living tissue. This toolkit allows us not just to observe the biological software, but to actively debug and reprogram it.

6.1 Illuminating the Invisible

For most of the twentieth century, bioelectricity had a personality problem. Everyone knew nerves and muscles were electrical: you could make a frog's leg twitch with a spark, record a heartbeat on a strip of paper, and watch neurons fire in milliseconds. But morphogenesis—the slow, collective business of making a face, a limb, an eye—looked chemical and mechanical. It smelled like gradients of morphogens, not the crisp click of a voltage trace. Part of the mismatch came from biology; a larger part came from the tools.

The classic instrument of electrophysiology is a glass microelectrode, drawn to a tip finer than a human hair, filled with salt solution, and connected to an amplifier that turns tiny currents into a readable signal. Done well, it is one of the most exquisite forms of measurement in science: you can impale a single cell and feel, electrically, whether it is healthy, excitable, or dying. The method is also brutally local. A microelectrode is a stethoscope pressed to one pixel of a living picture.

That locality was not a minor inconvenience; it shaped what counted as a plausible question. If you can measure one cell at a time, you tend to think in one cell at a time. The embryo becomes a mosaic of independent parts, each with its own voltage, its own story. The idea that there might be a coherent, tissue-scale *pattern* of membrane voltage—a landscape with hills and valleys spanning hundreds or thousands of cells—was hard to see, and therefore easy to dismiss.

Even when investigators suspected that something larger was happening, electrodes made it difficult to prove. Consider what it means to map a field with a single probe. You must decide where to measure, touch the tissue repeatedly, and hope the act of touching does not change what you are trying to measure. If the structure is moving, growing, or healing—as embryos cheerfully do—the coordinate system itself shifts under your hands. One can imagine the weary patience of early developmental electrophysiologists: hours at a microscope, trying to keep a tiny preparation alive and stable while taking point measurements that, at best, could be stitched into a crude map.

There is another, deeper limitation. A microelectrode reports a number: the difference in electrical potential between the inside of a cell and the outside bath, usually tens of millivolts. That number is meaningful, but it is not self-explanatory. In a neuron, you can interpret voltage changes as spikes. In an embryo, the important signals are often *resting* voltages—slow shifts that do not announce themselves with fireworks. They may matter not because a cell changes by 100 millivolts in a millisecond, but because a region stays slightly more negative than its neighbors for hours. A point measurement is a single temperature reading; morphogenesis cares about weather.

The first major step toward making bioelectric patterns visible was therefore not a new theory, but a new kind of eye. Voltage-sensitive dyes were invented for neurophysiology, where researchers wanted to watch activity spread across neural tissue. These molecules embed in the cell membrane (or associate with it) and change their fluorescence depending on voltage. Shine light of one color, and the dye emits light of another; the emitted brightness (or spectrum) shifts as the membrane potential shifts. With a camera, one can record not just a time-series from one cell, but a movie of voltage changes across a surface.

For morphogenesis, this change in vantage point is transformative. Development is not usually about rapid action potentials, but about stable domains: patches of tissue that are, on average, more depolarized or more hyperpolarized. A dye that can report these differences turns a hypothesis—"maybe there is a prepattern"—into an image you can point to. It also changes the emotional tone of the science. A jagged line on an oscilloscope is abstract; a glowing gradient on an embryo feels like a map. Scientists, like everyone else, are moved by what they can see.

A concrete example comes from work in *Xenopus laevis*, the African clawed frog, a stalwart of developmental biology because its embryos are large, robust, and tolerant of manipulation. In the early 2010s, Dany Adams, Rachel Masi, and colleagues in Michael Levin's lab used voltage-sensitive dyes to reveal a striking phenomenon: before the eyes are anatomically visible, before the familiar transcription-factor story has fully declared itself, a cluster of cells in the anterior region shows a distinctive bioelectric signature. This was not a fleeting flicker. It was a spatial domain of hyperpolarization that appeared as a coherent patch—a pre-eye-field electrical state.

The significance is not merely that such a patch exists, but that it can be *read* and *perturbed*. If you depolarize the relevant cell lineages during the critical window, eye development is disrupted. More provocatively, if you impose eye-like bioelectric states in tissue that does not ordinarily make an eye, you can coax ectopic eye structures in unusual locations. That kind of result is hard to obtain with purely descriptive measurements. It depends on being able to recognize an electrical state as a pattern, then treat it as a variable you can push around.

Voltage-sensitive dyes are not magic; they are chemistry with opinions. Some dyes respond to voltage because the electric field across the membrane shifts the dye's electronic structure; others redistribute across the membrane depending on voltage, altering local concentration. Many are sensitive not only to voltage, but to membrane composition, dye loading, illumination intensity, and photobleaching. The embryo is not a clean cuvette; it is a dynamic, light-scattering, protein-rich environment that tends to autofluoresce. A fluorescence map can therefore be an honest measurement and still be a treacherous one.

This is why the first generation of whole-field voltage imaging carried a particular scientific tension. The images were persuasive—sometimes hauntingly beautiful—but critics could reasonably ask whether the colors truly represented voltage. Calibration is difficult: an electrode provides an absolute millivolt value; a dye often provides relative changes in intensity. And relative to what? The tissue thickness differs from place to place; light penetrates unevenly; different cells may take up different amounts of dye. Some laboratories used ratiometric strategies (two wavelengths, one voltage-sensitive and one reference) to correct for loading and illumination. Others combined optical imaging with occasional electrode measurements to anchor the scale. The field learned, slowly, how to treat the fluorescent picture not as an illustration, but as a quantitative dataset with error bars.

The real conceptual shift, though, did not require perfect calibration. Morphogenesis often depends on *differences*—boundaries, contrasts, and domains—more than on an absolute number. If a region consistently appears more hyperpolarized than its surroundings, and if that region predicts where an organ will form, the pattern is already informative. It is like seeing that a coastline exists even if your ruler is imperfect. The key is consistency across embryos and across measurement methods.

Voltage-sensitive dyes also changed what one could do ethically and practically. Electrode impalement is invasive and can damage cells, especially small embryonic cells or delicate epithelia. Dyes, while not always benign (some are toxic at high concentration, and intense illumination can harm tissue), can be applied more gently and can report from many cells at once. That matters if you care about emergent properties—the collective behavior of a multicellular network coupled by gap junctions. A point measurement can tell you a cell's voltage; it cannot easily tell you how a tissue maintains a gradient, how a boundary sharpens, or how a domain migrates as the embryo elongates.

As cameras improved, the pace of the story quickened. Sensitive detectors, faster frame rates, and better optics made it possible to watch voltage patterns evolve in real time, not just as static snapshots. Developmental biologists could ask questions with a new grammar: *When* does an electrical boundary appear relative to a gene-expression boundary? Does it precede it, follow it, or oscillate with it? If you perturb the embryo mechanically or chemically, does the voltage pattern recover, drift, or switch to a new stable arrangement? These are the kinds of questions that make bioelectricity feel less like an ornament and more like a control layer.

Then came a second leap: genetically encoded voltage indicators, usually abbreviated as GEVIs. Instead of bathing the embryo in a chemical dye, one introduces DNA encoding a fluorescent protein engineered to change its fluorescence with membrane voltage. The protein is expressed by the cells themselves and targeted to the membrane. The idea sounds almost like science fiction the first time you hear it: a cell manufactures its own voltmeter and displays the readout in color.

GEVIs inherit much of optogenetics' appeal: specificity. With the right genetic control, you can express the indicator in particular tissues, cell types, or developmental stages. You can avoid staining everything—including the yolk, the extracellular matrix, and any inconvenient debris—and instead label exactly the cells whose voltages you care about. In principle, you can also maintain expression for long periods, enabling chronic imaging across days of development or regeneration.

The engineering behind GEVIs is subtle. Many designs fuse a voltage-sensing domain (often derived from proteins that naturally respond to voltage) to a fluorescent reporter whose brightness or spectrum depends on the domain's conformation. Some strategies use Förster resonance energy transfer (FRET), where energy passes from one fluorophore to another in a voltage-dependent manner, producing a ratiometric signal. Others use single fluorescent proteins whose brightness changes as the voltage sensor shifts. Each design involves trade-offs: sensitivity versus speed, brightness versus photostability, and spectral properties that must play nicely with other reporters and with the tissue's own autofluorescence.

For morphogenesis, speed is usually not the limiting factor. Embryonic patterning is slow compared to neural spiking; one can afford indicators that respond in tens of milliseconds if the biological signals evolve over minutes or hours. Brightness, phototoxicity, and stability become more pressing. You want to image for long periods without bleaching the signal or cooking the embryo with excitation light. Here, the recent development of far-red voltage indicators is more than a technical footnote. Longer wavelengths penetrate tissue better and tend to cause less photodamage, and they free up the green channel for other reporters (like calcium indicators or gene-expression markers). Voltage imaging, in other words, begins to join the crowded color palette of modern developmental microscopy rather than monopolizing it.

An anecdote from the history of physiology makes the conceptual distance clear. In the late eighteenth century, Luigi Galvani famously observed that frog legs could twitch when touched by metal, leading him to argue for "animal electricity." His experiments were crude by modern standards, but the lesson was enduring: the living body is electrically excitable. What Galvani could not do was see an electrical pattern spread across a developing organism. Even in the mid-twentieth century, when Hodgkin and Huxley could model the squid giant axon with mathematical elegance, the embryo remained electrically dim because its voltages were slow, distributed, and difficult to access. The current era's voltage imaging would have looked like a new sense.

That new sense alters not only measurement, but imagination. Once you can watch bioelectric patterns, you can start to treat them like other fields that biology already takes seriously: chemical gradients, mechanical stress maps, gene-expression atlases. A voltage map can be compared to a fate map. A voltage boundary can be compared to a compartment boundary. The question becomes: does the voltage pattern simply reflect the underlying developmental program, or does it participate in writing it?

There is a temptation, when confronted with glowing images, to slide too quickly into metaphor: to speak of electrical "thoughts" of an embryo, or to imagine a literal blueprint drawn in voltage. Metaphor is not the enemy—it often guides discovery—but voltage patterns are not messages in a human language. They are physical states that bias cell behavior through known mechanisms. A membrane potential can regulate the opening of voltage-gated calcium channels, altering calcium signaling. It can influence the transport of charged molecules through channels and transporters, changing intracellular pH or the distribution of small signaling molecules. Through gap junctions, it can synchronize or differentiate regions by allowing ions and small metabolites to pass from cell to cell. The map matters because it changes what cells do, not because it carries a secret code etched in lightning.

At the same time, there is a sober reason to take the "pattern" idea seriously: robustness. Embryos are remarkably good at reaching the correct anatomical outcome despite noise, variation, and injury. A distributed control layer—a tissue-scale electrical state that can be sensed, maintained, and repaired—is an appealing candidate for how robustness might be achieved. A single gene mutation can be catastrophic, but a network state can, in principle, be stabilized by feedback. If the voltage of a region drifts, ion channels and pumps can compensate; if a boundary blurs, coupling and differential conductances can sharpen it. Imaging is what makes such feedback plausible to study, because feedback is not a property of a point; it is a property of a system.

The craft of voltage imaging in embryos has therefore become a kind of negotiation between physics, biology, and humility. One must choose reporters carefully, control illumination, and correct for motion. One must worry about phototoxicity, about dye buffering, about whether expression of a membrane protein might itself perturb the membrane's electrical properties. One must also accept that the measurement is not the phenomenon. A fluorescent map is an inference, built on models of optics and chemistry, and it must be tested by perturbations and cross-validation.

Yet the payoff is difficult to overstate. With whole-field imaging, bioelectricity stops being a whisper you hear by poking at one cell and becomes a chorus you can watch. The embryo is no longer electrically silent between the dramatic events of nerve and muscle activity. Instead, it is electrically textured: domains, gradients, and boundaries that can precede anatomy, predict anatomy, and sometimes be steered.

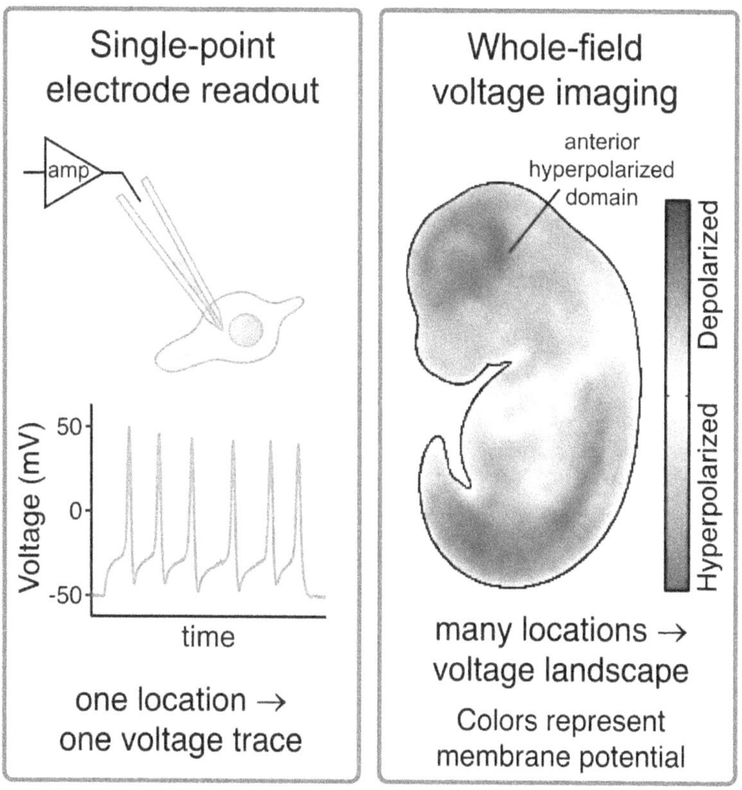

When you view an embryo as a voltage landscape, a new kind of causality becomes thinkable. A gene can be a switch, but a landscape can be a constraint: it can make certain cellular decisions easier and others harder, simply by biasing the physical conditions under which signaling and transport occur. The most compelling images are not the ones that look dramatic, but the ones that look *organized*—a crisp border where there was no anatomical border yet, a persistent patch that foreshadows an organ, a gradient that aligns with an emerging axis. Those are the moments when electricity stops being something the embryo *has* and becomes something it *uses*.

6.2 The Optogenetic Stylus

Seeing a voltage landscape is one thing; touching it is another. Developmental bioelectricity is full of patterns that look purposeful, but a picture alone cannot settle the harder question: are those patterns merely footprints left by other processes, or are they part of the steering wheel? The old way of pushing on voltage—pouring ions into the bath, blocking channels with drugs, swapping extracellular solutions—works, but it is like trying to edit a document by changing the room temperature. Everything changes at once, and the organism often compensates before you can learn what mattered.

Optogenetics offers a more surgical form of influence. The word is built from two simple ideas: *optical* control and *genetic* targeting. If a cell can be made to express a light-sensitive ion channel in its membrane, then a flash of light becomes an electrical instruction. The cell does not merely glow; it *obeys*.

The core trick came from microbes. Many single-celled organisms live in a world where light is information: it predicts where food might be, where danger might be, and which direction is up. Some algae and bacteria evolved proteins that change their shape when they absorb photons, and in doing so open a pathway for ions to cross the membrane. In a microbe, this can help steer movement toward or away from light. In an embryo or a regenerating tissue, the same protein becomes a controllable valve in the membrane's electrical economy.

To understand why this is powerful, it helps to keep one piece of membrane physics firmly in mind. Cells maintain ion gradients: sodium, potassium, chloride, calcium. Pumps and transporters spend metabolic energy to keep these ions unequally distributed across the membrane. That inequality is stored energy, like water behind a dam. Ion channels are the gates. When a gate opens, ions flow down their gradient, and the membrane potential—the voltage across the membrane—shifts. In most non-neural tissues, the membrane potential sits at a relatively stable "resting" value, not because nothing is happening, but because a balance of currents holds it there.

Optogenetic channels let you add a current that is timed and placed with unusual precision. A pulse of light can depolarize a targeted patch of cells (making the inside less negative) or hyperpolarize it (making the inside more negative), depending on the channel used and the ions available. The result is not a vague pharmacological nudge. It is closer to writing with a pen: localized, reversible, and capable of forming patterns.

Historically, the appeal of optogenetics was first felt most sharply in neuroscience, where electrical activity is fast and familiar. In 2005, Karl Deisseroth and colleagues helped establish that channelrhodopsin-2—a blue-light-gated cation channel from green algae—could be expressed in mammalian neurons to trigger action potentials with light. The cultural consequence was immediate: a new way to test what specific neurons do in behavior, not by lesioning tissue, but by switching defined cells on and off in living animals. If electrodes had been the stethoscopes of electrophysiology, optogenetics felt like remote control.

For morphogenesis, the stakes are different and, in some ways, stranger. The goal is not to evoke a spike, but to impose a *state*. You may want a group of cells to remain slightly hyperpolarized for hours because that state biases gene expression, cell migration, proliferation, or boundary formation. The time scale is slow, the feedback loops are numerous, and the tissue is not merely responding; it is building itself while you intervene.

This is why it is useful to think of optogenetics not as a set of gadgets but as a stylus: a tool for writing patterns into the bioelectric fabric of tissue. A stylus does not simply make noise; it makes shape. With the right light pattern, you can draw a stripe, a spot, or a gradient of depolarization across an epithelium. You can address one side of a regenerating fragment and not the other. You can vary the duty cycle of illumination—how long the light is on versus off—to tune the average voltage shift without constant exposure. You can do it again and again, with the same tissue, without adding new chemicals each time.

The library of optogenetic actuators expanded quickly beyond the first channelrhodopsins. Some channels conduct cations (positively charged ions), which tend to depolarize cells. Others conduct anions (negatively charged ions), particularly chloride, which often hyperpolarize cells by pulling the membrane potential toward the chloride equilibrium potential. Anion-conducting channelrhodopsins, such as GtACR1 from the cryptophyte alga *Guillardia theta*, became especially valuable because they can produce strong hyperpolarizing currents with relatively low light intensities. In neural applications, these are often described as "inhibitory" tools. In morphogenesis, the language of inhibition and excitation is less helpful than the language of voltage landscapes: you are raising and lowering regions of the electrical terrain.

Targeting is the other half of the power. Because the actuator is encoded genetically, expression can be confined to particular cell types using tissue-specific promoters, or mosaically expressed in subsets of cells. Developmental biologists have long been able to express fluorescent proteins in defined tissues; optogenetics extends that logic from passive labeling to active control. The difference is visceral. A fluorescent tag says, "watch me." An optogenetic actuator says, "move when I tell you."

The immediacy of optical control also forces a clearer view of what an "instruction" means in biology. A developmental system is not a simple circuit where a voltage pulse produces a predictable output. Cells adapt. They change their channel expression, alter gap junction coupling, and engage homeostatic mechanisms that pull voltage back toward a preferred range. This is not failure; it is life. It also means that a one-time voltage perturbation can be quickly absorbed, leaving little lasting effect.

This is where optogenetics begins to resemble control engineering. In engineering terms, a single pulse is an open-loop perturbation: you apply an input without measuring whether the desired state was achieved or maintained. What you often want is closed-loop control: measure the state, compare it to a target, and adjust the input to correct the error. Biology does this naturally all the time. Optogenetics makes it possible for the experimenter to do it too.

All-optical closed-loop schemes are now feasible because voltage can be read optically (with a GEVI) and written optically (with an optogenetic actuator). One can imagine a patch of tissue that is meant to remain hyperpolarized. A camera measures fluorescence from the indicator, software estimates whether the voltage is drifting, and patterned illumination nudges it back. In neuroscience, there are demonstrations of "optogenetic voltage clamp" ideas, where bidirectional actuators and fast voltage indicators are paired to hold a neuron near a desired voltage trajectory. Translating this to morphogenesis is not a simple matter of reusing the hardware; the tissue is larger, the dynamics are slower, and the relevant signals may be small. But the conceptual move is profound: the experimenter is no longer poking the system and watching it recover, but actively maintaining a chosen bioelectric pattern against the tissue's own tendency to normalize.

The possibility of bidirectional control strengthens this analogy. Some constructs are designed so that one wavelength depolarizes and another hyperpolarizes the same cells. BiPOLES and its variants are examples: engineered combinations of opsins that allow excitation and inhibition with different colors of light. In developmental contexts, such a tool is not merely convenient; it changes what patterns are possible. If you can only depolarize, you can paint bright regions onto a dark canvas. If you can also hyperpolarize, you can carve valleys and ridges with more finesse, shaping boundaries rather than simply flooding a region.

The concrete question, of course, is whether such electrical writing can change anatomy in a way that is not easily achieved by chemical means. Here, developmental bioelectricity offers some of its most provocative results. In *Xenopus*, manipulations of membrane voltage have been shown to influence eye development, including the appearance of eye-like structures in unusual locations when appropriate voltage states are imposed in non-eye tissues. Those experiments were not originally optogenetic; they used ion channel misexpression and pharmacological manipulations. Optogenetics, however, provides a route to do similar reprogramming with better temporal control and without permanently altering the tissue's channel complement. It makes the intervention feel less like genetic surgery and more like writing and erasing.

A particularly vivid way to appreciate optogenetics' promise is to imagine an embryo as a living tablet screen. With voltage imaging, you can see the pixels—each pixel a cell or a small patch of cells—

changing brightness with electrical state. With optogenetics, you can touch the screen, not with your finger but with light, and draw a shape. The embryo will not passively display your drawing; it will respond, compensate, and sometimes incorporate it into its own developmental logic. The "stylus" metaphor works because it captures both agency and resistance. You can write, but you are writing into a material that has its own grammar.

That grammar includes gap junctions, the protein channels that connect neighboring cells' cytoplasm, allowing ions and small molecules to pass directly from cell to cell. In a bioelectrically coupled tissue, a voltage change introduced into one set of cells can spread, blur, or sharpen depending on the network's connectivity. Optogenetics therefore becomes a way not only to perturb individual cells but to probe the tissue's coupling architecture. Illuminate a small patch and watch how the effect propagates. Does it remain local, suggesting weak coupling? Does it spread as a smooth gradient, suggesting strong coupling? Does it jump along specific pathways, revealing anisotropies in connectivity? One can learn about the tissue's electrical wiring diagram by writing a pattern and watching how it is rewritten by the tissue itself.

There is also an important practical constraint that shapes the style of optogenetic experiments in embryos: light does not penetrate indefinitely. Blue light, often used for channelrhodopsins, scatters strongly and is absorbed more than red light in tissue. That means deep structures are harder to control optically, especially in older embryos or thicker tissues. The field's response has been partly technological (developing red-shifted opsins and red or far-red indicators) and partly experimental (working with accessible stages, transparent organisms, or tissue explants). These constraints matter because they nudge the questions asked. You can more easily write on the surface than in the interior, so surface epithelia and early embryos become natural proving grounds for the method.

Phototoxicity adds another constraint. Living tissue does not enjoy being illuminated intensely for long periods, particularly at shorter wavelengths. Optogenetic protocols therefore often involve brief pulses or low-duty-cycle illumination, and careful choice of wavelengths that minimize harm. Again, this is not a minor technicality; it shapes what is possible. A stylus that burns the paper is not very useful. The best optogenetic writing is gentle enough that the tissue's response is a response to voltage, not a wound response to light.

All these constraints can make optogenetics sound like it belongs to a niche of laboratories with expensive microscopes and custom software. That is partly true. Yet the deeper reason optogenetics matters for morphogenesis is that it provides a clean conceptual experiment: can an imposed bioelectric pattern *instruct* anatomy?

Chemistry blurs instruction with collateral effects. If you bathe an embryo in a channel blocker, you change not only voltage but often pH, cell volume regulation, and signaling pathways in complicated ways. Optogenetics still has side effects—expressing an extra membrane protein is not neutral—but it offers a rare form of causality in living systems: you can impose a specific electrical change at a specific time in a specific place, and you can stop imposing it and see whether the system returns. You can test thresholds: does a small hyperpolarization suffice, or must it be sustained? Does timing matter more than magnitude? Can a transient pattern trigger a stable morphological change, suggesting that the tissue has internal memory?

This language of thresholds and memory is not borrowed from physics for rhetorical flair; it is what complex dynamical systems do. Development often behaves as if it has stable "set-points": outcomes toward which it gravitates. If bioelectric patterns participate in defining these set-points, then writing with light becomes a way to push the system from one basin of attraction to another. In plain terms, the tissue might be willing to build one structure or another, and voltage is part of what decides.

It is hard to read this without feeling both exhilaration and unease. Exhilaration, because the method hints at a future in which birth defects might be corrected not only by gene editing or pharmacology, but by temporary rewiring of tissue-level electrical states during critical windows. Unease, because the method makes it clear that anatomy is not merely a genetic fate but a controllable outcome, and control invites both medicine and misuse. The ethical landscape is beyond any single technique, but optogenetics sharpens the question by making intervention more precise and, therefore, more plausible.

On the bench, the immediate value is more modest and more beautiful: optogenetics turns bioelectricity into an experimentally writable variable. A developmental biologist can ask, with unusual clarity, whether a particular electrical boundary is necessary, sufficient, or neither. One can draw a boundary where none exists and see whether cells sort themselves accordingly. One can erase a boundary transiently and see whether the tissue restores it. One can create competing patterns and watch which one the tissue accepts.

The most telling moments often arrive not as dramatic transformations but as quiet demonstrations of competence. Illuminate a region, and the tissue responds in a way that is spatially organized rather than chaotic. The voltage change remains coherent across cells, suggesting strong coupling. A gene-expression marker shifts in a domain aligned with the imposed voltage. A migrating cell population changes course as if it has encountered a new border. The point is not that light is a magical morphogen, but that voltage can be treated as a controllable, instructive state variable, as real as a concentration gradient or a mechanical tension.

The embryo, under the stylus of light, reveals a surprising intimacy between physics and form. Photons strike a membrane protein; ions flow; the electrical state shifts; and a cell's decisions tilt, not because it has been forced, but because the conditions under which it computes its next move have been altered. The intervention is almost weightless, yet it can be decisive, like changing the slope of a surface so that a rolling ball chooses a different path.

6.3 The Silicon Simulator

Once you can watch voltage patterns and nudge them with light, an awkward truth becomes impossible to ignore: bioelectricity is not a single dial. It is a web of coupled dials, each one attached to the others by feedback loops that do not announce themselves until you try to intervene. Open one channel, and the cell may close another. Change a membrane potential, and you may change calcium entry, pH regulation, transporter activity, cytoskeletal dynamics, and gene expression. At the level of tissue, gap junctions spread and average signals, but not always smoothly; topology matters, and topology itself can change as cells divide, migrate, and differentiate.

Human intuition is good at linear chains—A causes B causes C. Morphogenesis is rarely polite enough to obey. It behaves more like a dynamical system: a network with multiple stable modes of behavior, capable of recovering from perturbations, sometimes capable of flipping into a different stable mode when pushed too far. In plain language, tissues can have *preferred* outcomes, and those preferences are not written in one molecule but distributed across many interacting processes.

The modern response to this complexity is neither resignation nor blind faith in computation. It is a practical partnership: use experiments to reveal the important players, use models to keep track of the consequences, then return to experiments with sharper questions. In developmental bioelectricity, this partnership has a particular flavor. Electrical states are inherently physical, and physics is unusually friendly to mathematics. If you know the ion concentrations, the channel conductances, the membrane capacitance, and the coupling between cells, you can in principle compute currents and voltages. That "in principle" hides the difficult part: in living tissues, many of those parameters are unknown, time-varying, and context-dependent. Still, the attraction is obvious. A good simulator can act like a wind tunnel for bioelectricity: you can test ideas cheaply, explore scenarios that are difficult to set up in vivo, and learn which interventions are likely to be decisive before you touch a living embryo.

There is a historical echo here. In the 1950s, the Hodgkin–Huxley equations turned the squid giant axon into a solvable problem, demonstrating that a handful of voltage-dependent conductances could explain the action potential. That achievement did not replace experiments; it made them more meaningful. It told physiologists what to measure, what to perturb, and what kinds of mechanisms were sufficient. Developmental bioelectricity seeks an analogous clarity, but the object is not a single axon. It is a tissue: a crowd of cells coupled into a collective electrical entity.

One influential computational platform for this is the BioElectric Tissue Simulation Engine, usually called BETSE. Its ambition is straightforward to state and hard to execute: model a multicellular tissue as a physical system in which ions move, charges accumulate, membranes behave like capacitors, pumps consume energy to maintain gradients, channels open and close, and gap junctions connect cells into an electrical network. Unlike simpler models that treat voltage as a variable that can be assigned arbitrarily, BETSE attempts to compute membrane potential from the underlying movement of ions and charge. That matters because many biological consequences of voltage depend on the ionic context: a depolarization caused by sodium influx is not identical to one caused by chloride efflux, and a tissue-level field depends on extracellular as well as intracellular conductivities.

The value of such a model is not that it produces a single "answer" for what an embryo will do. Biology is too rich, and our measurements are too incomplete, for that kind of certainty. The value is that it makes hidden assumptions visible. If you propose that a patch of hyperpolarization can emerge spontaneously in a tissue, a simulator forces you to specify how: which pumps are stronger there, which channels are expressed, how the cells are coupled, what the extracellular environment looks like. If your proposal only works when you set parameters to unrealistic values, you learn something important. If it works across a wide range of plausible parameters, you learn something else: the phenomenon may be robust, and therefore a credible candidate for a real control mechanism.

It also becomes possible to run experiments that are ethically or practically difficult in living organisms. You can "knock out" gap junction coupling by setting it to zero, then restore it gradually and see how patterns change. You can create tissues with different cellular geometries and ask how the shape of the tissue itself biases the voltage landscape. You can impose hypothetical optogenetic actuation patterns—a stripe of depolarization, a moving spot of hyperpolarization—and observe how a coupled network responds over time. In a living embryo, doing these systematically would be slow and noisy. In silico, you can do it hundreds of times and learn which variables matter most.

One of the most useful conceptual contributions of modeling is the language of attractors. An attractor is a stable state toward which a dynamical system tends to evolve. A marble dropped into a bowl rolls to the bottom; the bottom is an attractor. If the bowl has two dips, the marble may settle in one or the other depending on where you drop it and how hard you push. In morphogenesis, the "marble" is the tissue's overall state, and the "bowl" is shaped by gene regulation, biochemical signaling, mechanics, and bioelectric coupling. A stable anatomical outcome—a correctly patterned head–tail, a limb with the right polarity—can be thought of as an attractor in a high-dimensional state space.

This framing has a practical implication: small perturbations may not matter, because the system returns to the same attractor. Larger or differently timed perturbations may tip the system into a different attractor, leading to a different anatomical outcome. That helps explain a familiar developmental mystery: why embryos can shrug off many insults but occasionally switch to a dramatically different phenotype. It also clarifies why bioelectric interventions can sometimes have surprisingly persistent effects. If a voltage manipulation shifts the system into a different basin of attraction, the tissue's own dynamics may maintain the new state even after the original perturbation ends.

Computational models make this idea testable. They allow you to explore how changes in network topology—for example, altering gap junction connectivity—reshape the attractor landscape. They allow you to ask whether stable voltage patterns can function as a kind of memory, not in the metaphorical sense but in the dynamical sense: a state that persists because the network's feedback maintains it. This is one route toward understanding how non-neural tissues might store information about patterning decisions without relying solely on DNA sequence changes.

Modeling can also help unify what might otherwise look like a bag of unrelated observations. A voltage-sensitive dye shows a hyperpolarized domain in an embryo; optogenetics shows that altering voltage can bias a developmental outcome; pharmacology shows that certain ion channels matter. A computational framework can connect these by proposing a mechanism: perhaps a pump establishes a gradient, gap junctions spread it, and the resulting voltage influences a secondary messenger that controls gene expression. The model can then be used to predict what would happen if you changed the pump's activity, or if you disrupted coupling, or if you changed extracellular ion concentrations. Each prediction becomes an invitation for an experiment that is less like fishing and more like following a map.

The most interesting models in this space do not stop at physics. They try to integrate bioelectric dynamics with gene regulatory networks and biochemical reactions, acknowledging that voltage and gene expression are not separate worlds. Bioelectricity can regulate gene expression, and gene expression can regulate the channels and pumps that set bioelectricity. This creates feedback loops that can produce hysteresis: the system's state depends on its history. A cell might remain in a hyperpolarized state even after the original trigger is removed, because the hyperpolarization induced gene expression changes that reinforce the hyperpolarization. Hysteresis is one of the simplest mathematical forms of memory, and it is deeply relevant to development, which must make decisions that persist.

Integrated models of this kind are sometimes described as bioelectricity-integrated gene and reaction networks. The naming is less important than the idea: treat voltage as a first-class variable alongside transcription factors and morphogens. In practice, this means writing equations not only for gene expression and chemical concentrations, but also for membrane potentials and ionic currents, and then letting the coupled system evolve. Even when parameter values are uncertain, exploring the behavior of such systems can reveal qualitative possibilities: stable patterns, oscillations, waves, and switches that might correspond to biological phenomena.

Artificial intelligence enters this story in a different register. Physics-based simulators like BETSE are built from explicit assumptions about how ions and currents behave. AI, particularly machine learning, can help when the explicit assumptions are insufficient or the parameter space is too large to explore by hand. A machine learning model can act as a surrogate: trained on simulation outputs or experimental data, it can quickly predict outcomes of new parameter combinations. It can also help identify patterns in high-dimensional datasets that are difficult for humans to see.

Consider what modern voltage imaging produces: movies of fluorescence across tissues, potentially combined with gene-expression markers, cell tracking, and morphological measurements. The dataset is not a single curve; it is a stack of spatial-temporal fields. Extracting meaningful features—boundaries, domains, temporal sequences—is precisely the kind of problem at which machine learning excels. A neural network can be trained to recognize a prepattern that predicts a later anatomical feature, or to classify whether a tissue is likely to regenerate correctly based on early bioelectric dynamics. Even without claiming deep understanding, such a tool can be useful as a warning system or a design aid: "this intervention is likely to push the tissue toward outcome A rather than outcome B."

AI also invites a tantalizing possibility: *inverse* design. Forward modeling asks, "if I change these channels and pumps, what pattern emerges?" Inverse design asks, "if I want this pattern, what changes should I make?" In engineering, this is routine. In biology, it is difficult because the system is underdetermined: many different interventions might produce similar outcomes, and some interventions may be impractical or harmful. Still, combining a simulator with an optimization algorithm can begin to suggest candidate interventions—including optogenetic stimulation patterns—that are more targeted than trial-and-error approaches. One can imagine specifying a desired voltage landscape and asking the model to propose light patterns that could maintain it, given a particular optogenetic actuator and tissue geometry.

A sober caution is necessary here. Models can seduce. A colorful simulation can look as convincing as a microscope image, especially to the untrained eye, and even to trained eyes when the aesthetic matches expectations. The danger is not that models are wrong; it is that they can be right for the wrong reasons. A model with many tunable parameters can often be made to fit observed data without capturing the real mechanism. This is why good modeling in biology is less about producing a perfect fit and more about producing *constraints*: predictions that would be difficult to satisfy unless the underlying idea is correct.

The healthiest modeling culture therefore treats disagreement as information. If a simulator predicts that a particular gap junction perturbation should erase a voltage boundary, but the boundary persists in vivo, that is not merely a failure. It suggests either that the tissue has compensatory mechanisms not in the model, or that the measured boundary reflects a different variable than assumed, or that the topology is different. The model becomes a diagnostic, pointing to what we do not yet understand.

A concrete example of modeling's value is found in planarian regeneration, where fragments of a flatworm can regrow a complete animal with correct head–tail polarity. The biological feat implies global decision-making: a fragment must decide which end will become a head and which a tail, and it must coordinate that decision across tissues. Experimental work has implicated ion channels and gap junction communication in these patterning decisions, and computational framing has helped articulate what "decision" means in a distributed system. Rather than imagining a single master signal, one can model a coupled network with multiple stable polarity states. A perturbation to coupling or membrane properties can bias the system toward one stable polarity outcome or another. The language of attractors is not a poetic flourish here; it is a way of describing how a tissue can make a binary choice robustly.

The deeper intellectual payoff is that modeling pushes bioelectricity toward something like a theory of *collective* control. Neurons compute by exchanging spikes; tissues may compute, in a looser but still meaningful sense, by exchanging ions and small molecules through gap junctions, maintaining spatial patterns of voltage that bias cell behavior. Some modeling work has explicitly explored the idea that non-neural bioelectric networks can implement logic-like functions and pattern detection. Whether one wants to call this "computation" depends partly on taste, but the underlying claim is modest and testable: networked electrical dynamics can transform inputs into stable, functional outputs. Morphogenesis is, among other things, an input-output problem: genes and environment go in; anatomy comes out.

The phrase "silicon simulator" can sound cold next to the warm, wet intimacy of embryos. In practice, it is often the opposite. A good simulator makes you respect the tissue more, because it makes clear how much work the tissue is doing. Homeostasis is not a background feature; it is active control. Robustness is not a miracle; it is a consequence of feedback loops that can be explored, challenged, and sometimes harnessed. When models include membrane capacitance, electrodiffusion, and energy-consuming pumps, they remind you that maintaining a voltage landscape is not free. It is a physical achievement, paid for in ATP, coordinated across cells, and shaped by geometry.

The most compelling use of computation in this area is therefore not as an oracle but as a conversation partner. It can keep track of what the human mind cannot easily juggle: the coupled dynamics of many ions, many cells, many feedback loops. It can propose counterintuitive interventions. It can tell you that two manipulations that seem equivalent are not equivalent once ionic context and coupling are considered. It can suggest that a small change in network topology might matter more than a large change in a single channel's conductance.

There is a particular kind of satisfaction when a model earns its keep. You measure a voltage pattern in an embryo, perturb a component, and the pattern changes in a way that the simulator predicted *before* you did the experiment. Or you observe a puzzling recovery—the tissue restores a boundary after an insult—and the model reproduces the recovery only if you include a particular feedback loop, pointing you to a mechanism to test. The satisfaction is not that the computer is clever; it is that the tissue's logic becomes legible.

The phrase "hacking" can imply recklessness, but the best hacking is respectful: it learns the system's rules and works with them. In developmental bioelectricity, the silicon simulator is a way of learning those rules without exhausting living material and without being fooled by the mind's preference for simple stories. It helps transform the dream of electrical control into something closer to engineering: specify a target state, understand the constraints, design an intervention, and then let the organism show you which parts of your design it accepts and which parts it rewrites.

When that dialogue succeeds, it can feel as if a new kind of biology is taking shape—one in which anatomy is not only described and catalogued, but treated as the emergent output of networks that can be measured, modeled, and gently steered, with light on the bench and equations in the background like a second, quieter microscope.

Chapter 7

Designing New Life Forms

Standard biological dogma suggests that an organism's shape is irrevocably tied to its genome, but recent breakthroughs have shattered this constraint. By liberating cells from their evolutionary history, we can now collaborate with artificial intelligence to sculpture living materials into functional machines known as Xenobots. This exploration reveals that the "software" of life is not hardcoded in the nucleus, but dynamically writable in the electric networks between cells.

7.1 From Frog to Machine: The Xenobot Genesis

A strange kind of engineering begins with an act that would have been familiar to embryologists a century ago: taking living tissue from a developing animal and watching what it does when removed from its usual context. What changed in the Xenobot work was not the blade or the microscope, but the partner on the other side of the bench: a search algorithm that proposes body plans no human designer would naturally sketch, then a set of careful hands that assemble those plans from the most ordinary of biological materials.

The raw material was the African clawed frog, *Xenopus laevis*, a workhorse of developmental biology because its embryos are large, robust, and forgiving of manipulation. Long before anyone spoke about "living robots," *Xenopus* taught researchers how cells decide what they are going to become, how tissues fold and fuse, and how embryos recover from perturbations. In the 1930s and 1940s, Hans Spemann's tradition of microsurgery on amphibian embryos made it clear that pieces of an embryo can reorganize in surprising ways when transplanted. Later, in the mid-twentieth century, the first tissue culture experiments revealed another unsettling fact: cells often keep their basic competencies even when removed from the body, as if they carry a repertoire of behaviors waiting for the right cues. The Xenobot story leans on that older knowledge but puts it to a new use: instead of asking what tissues *naturally* become, it asks what else they *could* become if assembled into unfamiliar geometries.

"Living robot" is a phrase that can mislead if it conjures metal joints and silicon chips, merely coated in biology. The early Xenobots, first reported by Michael Levin and Douglas Blackiston's group at Tufts University together with Joshua Bongard's group at the University of Vermont (2020), were entirely biological: clusters of frog cells, built without genetic modification and without any rigid scaffold. They were not miniature frogs, nor were they conventional machines. They were better described as *reconfigurable organisms*: living tissues arranged into a new body plan that exploits the cells' own forces, adhesions, and rhythms to do something observable and repeatable in the world.

Two cell types did most of the work. One was ordinary skin (epithelial) tissue from the frog embryo, which forms a protective surface and tends to seal into smooth boundaries. The other was cardiac muscle, which has a built-in motor: it contracts spontaneously and rhythmically. Anyone who has watched a tiny patch of embryonic heart tissue beat in a dish recognizes the eerie autonomy of that motion. It is "automatic" in the strict physiological sense: the tissue generates its own electrical activity and mechanical contractions without needing a brain to tell it what to do. The Xenobot design challenge was simple to state: can those contractions be arranged so that a cluster of cells moves across a surface in a controllable way?

A human engineer could attempt a design by intuition: place the muscle on one side to push, add a notch to create asymmetry, hope friction does the rest. But intuition is notoriously unreliable in soft matter, and living tissues are the softest matter of all: they deform, heal, and change their internal organization while you watch. This is exactly the kind of setting where "design" starts to look less like drawing a blueprint and more like selecting from a vast space of possibilities.

That is where evolutionary algorithms enter. The phrase can sound grandiose, but the core idea is disarmingly concrete. Suppose you can represent a candidate Xenobot body as a simple digital object—a small 3D arrangement of blocks, some designated as contractile tissue and others as passive tissue. You can then simulate how that object would behave under simplified physics: how it deforms, how it pushes on a surface, whether it drifts in a straight line or spins in place. If you can score each candidate by a goal (move forward efficiently, turn predictably, corral particles), you can use an algorithm that mimics natural selection: generate many variants, keep the better performers, mutate and recombine them, and repeat for thousands of rounds. The computer is not *discovering* life's secrets in any mystical sense; it is doing something closer to an audition, trying out forms at a speed no laboratory could match.

The crucial feature is that the algorithm's search space was not "genes" but *morphology*: geometry, placement of tissues, and overall shape. In other words, it treated anatomy as a programmable variable.

A subtle philosophical shift hides inside that technical choice. If you assume the genome fully specifies the organism, then novel behavior should require genetic alteration. If you allow that living tissues have a kind of latent versatility—that their collective dynamics can be redirected by new arrangements—then you can try to "program" by reshaping, not rewriting DNA.

A persistent problem in robotics and AI is the "sim-to-real gap": systems that perform beautifully in simulation often fail in the physical world because the simulation neglected some detail that turns out to matter. Living tissue magnifies this risk. Cells are not standardized parts manufactured to tight tolerances. A clump of cells may be slightly more adhesive than expected; a muscle patch may contract with a different amplitude; a surface may be a little wetter or rougher. The Xenobot pipeline confronted this gap directly by narrowing the representation to what could plausibly be built and by filtering candidate designs for robustness. The point was not to predict every molecule. It was to find shapes whose basic function survives the messiness of biology.

Just after the algorithmic search had produced a set of candidate body plans, the work moved from computation to craft. Under a microscope, embryonic tissues were dissected and assembled with fine tools—often compared, with only slight exaggeration, to tiny forceps and a hair-thin needle. The assembly step matters because it is the moment where "design" becomes literal. The designer is no longer nature alone, and no longer a human sculptor guided by intuition, but an iterative partnership: algorithm suggests, biologist builds, organism acts, results feed back into improved constraints and models.

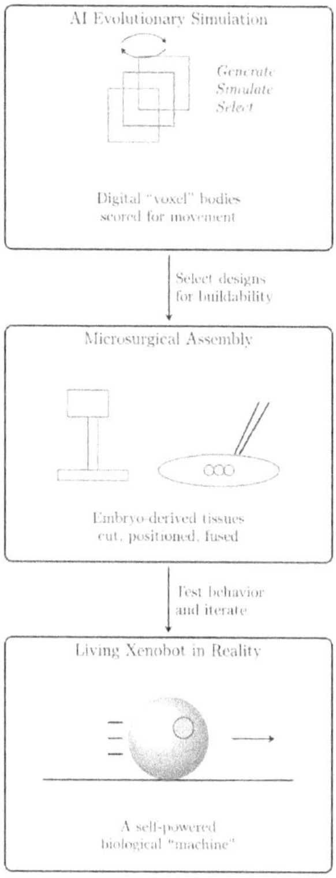

The assembled Xenobots were small, typically on the scale of a millimeter, visible as tiny pale specks that could nonetheless traverse a dish. Their motion was not the smooth glide of a fish or the coordinated stepping of an insect. It was a kind of pulsing creep or swim, driven by the contraction of the cardiac tissue. If the muscle was placed symmetrically, the organism might jitter in place. If offset, it might move forward. Some shapes traced arcs; others rotated. What mattered was that the motion was produced by *endogenous* forces: the bot carried its own actuator, fueled by the chemical energy already stored in its cells.

This is the point where the word "robot" begins to earn its keep, if we use it in a functional rather than material sense. A robot, in everyday language, is something built to do work in the world, with behavior that is at least partly predictable from its design. Xenobots were built, and they did work: they moved. Yet they were also not machines in the usual sense because their components were not passive. Skin cells do not merely "cover"; they secrete, adhere, and heal. Muscle cells do not merely "contract"; they are excitable tissues whose rhythm emerges from ion channels and membrane potentials. Even when removed from the embryo, they remain part of a deeply evolved physiological repertoire.

That repertoire makes Xenobots interesting in a way that can be missed if one focuses only on the spectacle of a "living robot." Traditional machines are brittle: damage a gear tooth and the system fails. Living tissues can be robust because they maintain themselves. A cut in a cell cluster can seal; a disruption can be absorbed by rearrangement. This does not mean Xenobots are indestructible. They have limits, and they run down as their internal stores are depleted. But the basic contrast matters: the design target is not a rigid mechanism but a self-organizing material.

A helpful historical analogy comes from the earliest decades of cybernetics and robotics, when researchers built simple machines that seemed to exhibit lifelike behaviors without any elaborate intelligence inside. In the late 1940s and early 1950s, the neurophysiologist W. Grey Walter built small wheeled devices—his "tortoises"—with a handful of electronic components. They could seek light, avoid obstacles, and appear to hesitate or "decide," all from feedback loops that linked sensors to motors. Walter's point was not that the machines were conscious, but that purposeful-looking behavior can emerge from simple rules embedded in a body interacting with an environment. Xenobots reprise that lesson in a biological key. Their "controller" is not software in a microprocessor. It is the physics of deformable tissue coupled to the electrophysiology of excitable cells, producing behavior that is both constrained and, in detail, somewhat unpredictable.

The evolutionary algorithm adds a twist Walter did not have: it searches for body plans that exploit those feedbacks. When a simulation scores a design as a good mover, it is effectively saying: *this arrangement of contractile and passive tissue is likely to harness the medium in a useful way*. The microsurgeon then instantiates that hypothesis in living matter. When it works, the result feels like a new kind of causality: not "genes cause form which causes function," but "form causes function even with the same genes."

A skeptic might object that this is still biology as usual: heart muscle beats, skin holds together, and the cluster moves because any asymmetry would produce some net push. The more interesting response is that the performance depends on the *specific* evolved geometry, and that controls can test this dependency. In the 2020 work, altering the layouts—for example by changing orientations or disrupting the evolved placement of tissues—reduced performance. That kind of control matters because it shifts the interpretation from "any random clump might wriggle" to "a designed arrangement reliably yields a behavior." Design, here, is not merely the act of cutting and pasting; it is the deliberate selection of a configuration from many alternatives.

This is also where the bioelectric theme quietly exerts pressure. Cardiac cells beat because their membranes maintain voltage differences, governed by ion channels and pumps, and because those electrical states coordinate across tissue. Even in a tiny muscle patch, electrophysiology gives rise to rhythm and synchronization. The Xenobot does not need a nervous system to convert energy into movement; it needs a collective of cells whose electrical and mechanical dynamics can settle into a stable pattern of activity. When those cells are embedded in a new geometry, the pattern produces a new outcome. The "software" is not a line of code but a distributed, excitable medium that can be steered by architecture.

It is tempting to treat the algorithmic design step as the true innovation and the microsurgery as mere fabrication. That would miss something essential. Building Xenobots is not like 3D printing a plastic part. Living tissues resist standardization, and the act of assembly is itself an intervention into a dynamic system. When two tissue fragments are brought into contact, they fuse, exchange signals, and reorganize. Boundaries smooth; tensions redistribute. The builder is not assembling inert bricks but initiating a process that continues after the tools are withdrawn. The design is therefore partly explicit (shape, placement) and partly implicit (the tissue's own self-organization).

Seen from this angle, the Xenobot pipeline resembles an older pattern in biology: the interplay between *formal* models and *material* experiments. D'Arcy Thompson, writing in 1917, argued that biological form is constrained by physical law, and he drew transformations and geometrical mappings as if an organism were a pliable diagram. Thompson did not have computation, and he did not imagine building organisms to specification, but he framed anatomy as something that can be reasoned about in terms of forces and constraints. The Xenobot work makes that reasoning operational: propose a shape under a physical model, test it in living matter, and revise.

The broader implication is not that frogs are secretly robots, or that intelligence has been smuggled into tissue. It is that living cells are a programmable substrate in a sense that is easy to underestimate. "Programmable" does not necessarily mean "fully predictable." It means that the system has degrees of freedom that can be deliberately set, and that setting them can redirect the system into qualitatively different behaviors. In classical engineering, degrees of freedom are often mechanical: lengths, angles, stiffness. In living engineering, degrees of freedom include geometry, tissue composition, and the physiological states that cells bring with them.

That combination also forces a reassessment of what counts as a "machine part." In a mechanical device, a motor's job is to produce force; it is not also a sensor, a repair technician, and a chemical factory. In living tissue, those roles are entangled. A patch of cardiac muscle is not merely an actuator: its rhythm responds to its electrical environment, its coupling to neighbors, and its mechanical load. The boundary between "control" and "body" blurs. The body *is* part of the control system, and the algorithm can exploit that by searching body plans rather than control policies.

The first Xenobots did not solve practical problems in medicine or environmental cleanup. Their achievement was more basic and, in its own way, more unsettling: they demonstrated a working bridge from abstract search in a virtual world to a new, functional anatomy in the real one, built from unedited cells. That bridge matters because it offers a recipe that can be generalized. If an algorithm can propose shapes that harness contractile tissue for locomotion, it can, in principle, propose shapes that harness other tissue competencies for other tasks: pumping fluid, gripping, sorting particles, or interacting with chemical gradients. And because the building blocks are biological, the resulting machines inherit biological properties by default: softness, compatibility with aqueous environments, and the ability to degrade rather than persist as plastic debris.

A final, easily overlooked detail makes the Xenobot genesis feel less like science fiction and more like developmental biology turned outward. The cells used in these constructs were not taught a new trick by gene editing. They were doing what cells already know how to do: stick to each other, form an outer layer, contract rhythmically, maintain physiological coherence. The novelty lay in the *context*—a context supplied by a design algorithm that does not share the embryo's assumptions about what a body should look like. When the construct begins to move, the impression is not that life has been manufactured from scratch, but that life has been persuaded, by geometry and constraint, to reveal a hidden range of possibilities.

7.2 Software Without Silicon: Programming the Swarm

A Xenobot built from frog cells carries no new genes. Nothing has been edited, inserted, or deleted. Under a microscope, the cells look like what they are: ordinary tissues from an ordinary animal. And yet the behavior can feel like an argument with common sense. When people hear "programming," they tend to picture a line of code, a chip, a designer who specifies each step. In living systems, the surprise is that a great deal of "software" is not written in DNA as a fixed script and not stored in neurons as thoughts. It lives in the rules by which cells coordinate: what they adhere to, what they pull on, what they secrete, and—crucially for morphogenesis—what electrical states they share.

The phrase *software without silicon* is not a metaphor for mere complexity. It names a specific kind of control: distributed, embodied, and often invisible unless you know how to look for it. A multicellular group can compute in the way an ant colony computes: no single ant "knows" the plan, yet trails form, nests adapt, and food is allocated. The Xenobot work took that idea and made it tactile. If you can rearrange a small set of tissues into a new geometry, you can coax a cell collective into behaviors that are not part of the frog's normal repertoire—including a peculiar kind of reproduction that depends on motion rather than growth.

The most famous example is *kinematic self-replication*, reported by Sam Kriegman, Douglas Blackiston, and colleagues (2021). "Kinematic" simply means movement-based. This replication does not resemble cell division, budding, or any familiar developmental program. Instead, a Xenobot acts like a tiny bulldozer: it moves through an environment containing loose cells, nudging and corralling them into a pile. Under the right conditions, that pile compacts, its cells adhere, and a new Xenobot-like cluster forms. The offspring then swims away and can, for a limited number of generations, repeat the process.

It is hard to overstate how conceptually disruptive that is. Reproduction, in schoolbook biology, feels like the most genetically determined of all traits: DNA is copied; an embryo develops; a body plan unfolds. Kinematic replication shows something different. The genome provides the components and the basic cellular skills—adhesion, motility, excitability, the ability to form an epithelium—but the reproductive *strategy* can arise from a novel ecology and a novel body shape. The "instruction" to replicate is not a gene that says "replicate like this." It is an emergent consequence of the bot's geometry, the physical properties of the cells, and the availability of free building blocks in the environment.

A historical echo helps here. In 1922, the mathematician and biologist D'Arcy Thompson remarked that many forms in biology can be understood as the outcome of forces rather than as the direct expression of a blueprint. That view was, for decades, a minority taste compared with the romance of genetic instruction. Then, in 1952, Alan Turing published his paper on "The Chemical Basis of Morphogenesis," proposing that simple reaction-diffusion systems could generate complex patterns such as stripes and spots. Turing's point was not about frogs or robots; it was about a class of explanations where pattern appears because the system has an instability that amplifies small differences into organized structure. Xenobots push a related thought into the realm of behavior and reproduction: a new pattern of life can appear because a cell collective, placed into a new physical configuration, expresses capacities that were latent all along.

Kinematic replication also depends on a subtle shift in what counts as an "individual." A frog embryo is a single organism in a familiar sense: one genome, one body, one developmental trajectory. A Xenobot is a temporary coalition of cells cut from an embryo and persuaded into a new unity by adhesion and self-organization. It is an individual not because it has a unique genetic identity but because it has a coherent boundary, a stable set of internal interactions, and a persistent behavior. If one Xenobot can collect free cells and turn them into another coherent unit, then individuality has become, at least partly, a property of organization rather than lineage.

This does not mean Xenobots are alive in some exotic new way. They are alive in a very old way: they are made of living cells. The novelty lies in how those cells coordinate, and that takes us back to bioelectricity.

Every cell has an electrical state across its membrane: the inside and outside carry different concentrations of ions, so there is a voltage difference called the membrane potential, often written as V_m. Neurons make the headlines because they produce fast spikes, but most cells are electrical. Skin cells, muscle cells, and stem cells all use ion channels and pumps to maintain V_m. When groups of cells are connected by gap junctions—tiny protein channels that directly connect the cytoplasm of neighboring cells—they can share ions and small signaling molecules. A tissue can then behave less like a pile of independent units and more like a network, capable of spreading electrical states and coordinating responses over distances that are large compared to a single cell.

Michael Levin's lab and others have argued, across many experiments and over many years, that such bioelectric networks act as an instructive layer for morphogenesis: spatial patterns of V_m can bias gene expression, influence cell proliferation and migration, and help maintain a "target" anatomy in regeneration. Even if one does not accept the strongest version of the "bioelectric code" idea, the basic physiological facts remain: tissues have distributed electrical states, those states can be mapped and perturbed, and perturbations can have large downstream effects on form.

Xenobots offer a vivid stage on which to watch that layer at work, partly because they strip away the usual interpretive crutches. In an embryo, any one observation can be explained away as a consequence of "the program"—as if the genes must have intended it. In a Xenobot, the arrangement is artificial and the genome unchanged, so when behavior changes dramatically with shape, the argument shifts. Something other than genetic sequence is carrying the relevant information.

That "something" is not a single mechanism. It is a bundle of constraints and couplings. The physical side is immediate: a particular geometry makes it easier to push a loose cell cluster without breaking it apart; a notch can trap cells like a scoop; the curvature of a surface changes friction and flow. But the physiological side is equally important. A cluster of cells has to remain coherent long enough to become a new bot. Coherence is not merely glue. It involves cells recognizing neighbors, maintaining boundary integrity, and establishing stable internal states. Bioelectric coupling through gap junctions is one way tissues achieve that stability. A cohesive electrical network can help synchronize activity, coordinate contraction, and maintain differentiated states.

The kinematic replication experiments made this concrete by showing that morphology is not just the shell around a living system; it is part of the control system. The most effective replicators had a shape often described as "Pac-Man": a roughly circular body with a wedge-like mouth. As the bot moved, the mouth region corralled loose cells into a compact cluster. A nearly spherical bot, by contrast, tended to scatter cells rather than gather them. This is not surprising in hindsight—anyone who has tried to sweep crumbs with a round object knows the problem—but the larger point is that a new reproductive behavior could be *searched for* and *selected* by altering shape, without any molecular rewriting.

There is a deep and uncomfortable implication here. If reproduction can be reframed as an emergent behavior of a cell collective, then it becomes possible to imagine a spectrum between "non-replicating machine" and "self-propagating organism" that is not cleanly divided by genetics. The boundary becomes ecological: what materials are available, what constraints are present, what dynamics the collective settles into. That is why the kinematic replication results were reported alongside careful caveats. Under the described laboratory conditions, replication occurred for limited rounds and then halted. The system was not an endlessly self-propagating entity. It was a demonstration of possibility, not a release of a new species.

Swarm behavior provides a second window into software without silicon. Put a single Xenobot in a dish and it looks like a curious toy. Put many together and you begin to see collective dynamics: clustering, transient alignment, interactions that depend on collisions and local cues. In later work that broadened the platform beyond the earliest muscle-driven bots, xenobot-like constructs could self-assemble into cilia-driven movers—using tiny hairlike structures (cilia) that many epithelial cells naturally grow and beat in coordinated waves. Cilia are another example of living actuation that arrives "for free" with the cell's own developmental repertoire. Once you have dozens or hundreds of such agents, the interesting question becomes less "how does one move?" and more "what happens when many move together?"

The question is familiar to anyone who has watched starlings flock or fish school. Collective behavior can look choreographed, yet it can arise from simple local interactions: keep a certain distance, align direction, avoid collisions. Robotics has long borrowed from these biological examples. "Swarm robotics" designs groups of relatively simple robots that achieve a task by coordination rather than by central control. The twist in the Xenobot context is that the agents are not manufactured devices but living tissues with their own internal states.

A useful anecdote comes from the early days of collective behavior studies in social insects. In the 1980s and 1990s, experiments showed that ants can sort their dead, their brood, or their food into piles without any ant having a global plan. Each ant follows a local rule: pick up an item when it is isolated; drop it when similar items are nearby. From that, piles form. The "program" is not a blueprint; it is a rule that exploits statistics. Xenobot swarms invite a similar mode of thinking. A designer might not micromanage each bot; instead, one shapes the conditions—density, obstacles, loose cell availability, boundary constraints—so that local interactions produce an outcome.

In living swarms, bioelectricity adds a further layer because local interactions are not only mechanical. Cells can sense chemical gradients, electrical fields, and the presence of neighbors through direct contact. Gap junction coupling, when present within each agent, can integrate information across its body, affecting how it responds to its environment. Even without long-range coupling between separate bots, each bot is an electrical network, and the swarm is an ecology of networks.

Thinking this way can make the word "programming" feel more honest. A programmer does not always specify outcomes line by line. Sometimes the programmer specifies constraints and lets a system find a solution. That is exactly what happens in machine learning, where a loss function and a training environment shape behavior. Xenobot work takes a similar approach but uses biology as the substrate. The "loss function" might be forward movement, or the ability to gather loose cells. The "training" might be a simulated evolutionary search. The "deployment" is a living construct whose behavior is partly predicted and partly discovered at runtime, so to speak, when it meets the physical world.

This is also where humility is necessary. The temptation is to declare that life is simply a programmable material and that bioelectricity is the master dial. Real tissues are not obedient. They have their own attractors—stable patterns of organization that they tend to return to after disturbance. In developmental biology, those attractors are why embryos can self-correct. In synthetic morphology, those attractors are both a resource and a constraint. If you push a cell collective too far, it may reorganize into something else, or fail to remain stable. If you push in the right direction, the same self-correction that rescues an embryo can help a designed construct maintain its function.

There is a practical engineering lesson here that mirrors a lesson in regenerative medicine. When surgeons transplant tissue or when engineers seed cells into a scaffold, they often discover that the tissue does not simply fill the space as commanded. It remodels, contracts, vascularizes, and sometimes rejects the intended geometry. Successful tissue engineering increasingly relies on working *with* self-organization: providing cues and constraints that guide, rather than dictate, the final form. Xenobots make that lesson visible at a small scale. They behave less like machines assembled from dead parts and more like a coordinated crowd: capable, adaptive, and occasionally stubborn.

The philosophical stakes show up in a deceptively simple question: where is the "plan" stored? DNA stores sequences. Neural networks store synaptic weights. But a Xenobot's effective plan—how it moves, how it gathers cells, whether it can replicate kinematically—is distributed across morphology, mechanics, and physiology. The plan is partly in the shape, partly in the material properties of tissues, and partly in the electrical coupling that makes the tissue a coherent unit. In that sense, the Xenobot is not merely a designed object; it is a designed *process*.

Kinematic replication, in particular, forces a rethinking of what reproduction requires. It requires not only information and copying but also a way to gather and organize matter. Cells already know how to organize matter at the microscale. When given the right macroscale geometry and a supply of loose cells, they can be enlisted into an unfamiliar reproductive loop. The loop is awkward, constrained, and dependent on laboratory staging, but it is real enough to unsettle the idea that reproduction is a singular, genetically hardwired trick.

The most honest way to hold the Xenobot results in mind is to see them as an experiment on the boundaries of competence. A frog cell, in its usual life, contributes to a frog. Out of context, it can contribute to a mover, a collector, a temporary individual with a behavior that looks goal-directed. The genome has not changed, but the space of possible outcomes has expanded because the collective dynamics have been placed in a new basin—a new region of the system's landscape where different stable behaviors are accessible.

That landscape is not infinite. It is shaped by evolution. The very reason frog tissues can be repurposed is that evolution produced robust modules: epithelia that seal, muscles that contract, cilia that beat, cells that adhere and communicate. Xenobots do not escape biology; they exploit biology's deep regularities. What they add is a way to search those regularities systematically, treating morphology and physiology as knobs that can be turned to reveal new functions.

When a kinematic replicator nudges a handful of loose cells into a tight cluster and the cluster later swims away as a new coherent agent, the scene is both simple and uncanny. No DNA has been copied in the moment, no embryo has been gestated, no parent has "intended" anything. A piece of living matter has used motion to make another piece of living matter, and that is enough to make reproduction feel less like a sacred boundary and more like a continuum of strategies available to organized collectives.

7.3 Beyond Robots: The Future of Living Architecture

Calling a Xenobot a "robot" is useful at the moment you need to snap the mind out of its usual categories. It forces the question: if an object is made of living cells, repairs itself to a degree, and performs a designed task, why should it be excluded from engineering? But the word can also trap the imagination in the wrong place. A robot suggests a gadget. A gadget suggests a bounded function, like a toy that swims until its batteries die. The more provocative possibility is not a menagerie of microscopic living machines skittering across petri dishes, but a new approach to *construction* in biology: persuading cell collectives to build and maintain complex anatomy the way they already build embryos and heal wounds.

Tissue engineering began, in the late twentieth century, with a straightforward dream: treat cells as building blocks, provide them with a scaffold shaped like the desired organ, and let them fill in the structure. The scaffold could be collagen, a synthetic polymer, a decellularized organ, or a 3D-printed lattice. If the shape is right and the biochemistry is right, the cells should populate it and, ideally, the body will accept it.

That idea has scored genuine successes. Skin substitutes for burns, engineered cartilage, and certain kinds of engineered bladder tissue have reached clinics. Yet the dream of printing or scaffolding a fully functional kidney, heart, or liver has resisted brute force. The reasons are not mysterious. Organs are not just lumps of cells in the right geometry. They are living architectures with multiple cell types arranged in precise, interlocking patterns, threaded with blood vessels, permeated by nerves, and maintained by constant remodeling. A kidney is a forest of microscopic tubules arranged to exploit gradients; a heart is a spiral of muscle fibers tuned for coordinated contraction. Even if a scaffold could be printed with perfect geometry, the living tissue has to *stay* organized while it grows, adapts, and repairs itself.

The phrase *living architecture* names that deeper requirement. Architecture is not mere shape; it is structure that supports function over time. In living systems, "over time" includes development, maintenance, and response to injury. The provocative thought behind synthetic morphology is that cells already possess the competencies required for living architecture. The embryo reliably builds organs, vascularizes them, innervates them, and integrates them into a whole. A salamander regrows a limb with bones, muscles, and skin in the right arrangement. Human skin heals wounds by closing, remodeling, and restoring barriers. These feats are not achieved by laying bricks into place. They are achieved by coordinated collective decision-making, mediated by chemical signals, mechanics, and electrical networks.

Bioelectricity becomes relevant precisely because it is a medium for coordination at the tissue scale. A single cell's membrane potential, V_m, is a local state. But when cells are coupled by gap junctions, a tissue can maintain spatial patterns of electrical state, like a landscape of voltages. Those patterns can be stable, can change in response to perturbation, and can influence downstream gene expression and cell behavior. In this view, an organ is not built solely by instructing each cell which proteins to express. It is built by guiding a collective toward a target configuration, the way a thermostat guides a room toward a target temperature without specifying the micro-motions of each air molecule.

A historical anecdote captures why this shift matters. In the 1960s and 1970s, as organ transplantation became clinically viable, surgeons learned that anatomy is not enough. A transplanted organ must connect, integrate, and maintain itself within a new physiological context. The technical drama of the first heart transplants in 1967 was not merely surgical skill; it was immunology, physiology, and the problem of long-term integration. Tissue engineering inherited a similar lesson: fabricating an object that resembles an organ is not the same as creating a living system that can remain an organ in a body. The second problem is harder because it is about *control* over time.

The living-robot experiments offer a small-scale parable of that control problem. A Xenobot that moves is interesting, but a Xenobot that maintains coherence in an unfamiliar geometry is more revealing. The tissue does not fall apart at once; it seals, reorganizes, and continues. That persistence hints at a general principle: biological materials carry their own repair and maintenance routines. The engineering challenge is to harness those routines toward a desired target rather than fighting them with rigid scaffolds and external micromanagement.

Two contrasting approaches to building tissue make the difference vivid.

One approach is bottom-up construction: place cells where you want them, perhaps using bioprinting, microfluidics, or patterned scaffolds. This is an extension of traditional manufacturing, trading steel and plastic for collagen gels and living cells. The advantage is obvious: you can specify geometry with mechanical precision. The disadvantage emerges over days and weeks. Cells migrate. They contract and remodel the matrix. They respond to gradients you did not anticipate. If vascularization is insufficient, the core becomes hypoxic and dies. The printed geometry becomes a starting guess rather than a final product.

A second approach is top-down specification: rather than placing every cell, provide global cues that define a target state and rely on the tissue's own self-organization to achieve it. In the bioelectric framing, one of the most powerful global cues is a pattern of V_{m} and coupling: a physiological "setpoint" that biases cells toward particular fates, boundaries, and growth behaviors. The hope is not to eliminate scaffolds or biochemical factors, but to shift the emphasis from mechanical imposition to informational guidance.

This is not science fiction; it is a reorientation of a long-standing practice. Developmental biologists have always used "global cues." A gradient of a signaling molecule can position a limb. Mechanical tension can orient cell divisions. Electrical fields can influence wound healing, an effect known for over a century and often discussed under the heading of galvanotaxis or electrotaxis: cells can migrate in response to electric fields. What is new is the idea that endogenous bioelectric patterns can be deliberately read and written as a control layer for organ-level outcomes, and that this layer might be engineered with the same seriousness as gene regulatory networks.

That seriousness is deserved partly because electrical state is both upstream and downstream. Ion channels and pumps set V_{m}, and V_{m} influences gene expression, proliferation, migration, and differentiation. But gene expression also changes the repertoire of ion channels.

The system is a loop, not a hierarchy. In loops, control is not about a single master instruction; it is about stabilizing desirable attractors and destabilizing undesirable ones. This language, borrowed from dynamical systems, can sound abstract, yet it maps cleanly onto what patients care about. A scar is a stable but undesirable attractor of wound healing. A regenerating limb is a desirable attractor. Cancer can be seen, in part, as a tissue that has fallen into a pathological attractor, maintaining growth and invasion despite the body's constraints.

Living architecture, then, is not merely about growing replacement parts. It is also about *reprogramming* tissues already in the body. Imagine telling cells not only "become cartilage" but "assemble into a joint surface of this curvature, integrate with this bone, maintain this lubrication." That is the level of specificity organs require, and it is difficult to achieve with molecular cues alone because the relationship between molecular state and organ geometry is not one-to-one. Two tissues can express similar sets of genes yet have different shapes because they occupy different positions in a tissue-level decision-making network. Bioelectric patterns are plausible candidates for carrying some of that positional and goal-related information.

The xenobot results also broaden the horizon beyond organs as static objects. "Regenerative medicine" often conjures replacement: a new kidney, a patched heart. But living architecture includes dynamics: coordinated contraction, rhythmic secretion, adaptive remodeling in response to load. Consider the heart. A scaffold shaped like a ventricle is not enough; the tissue must conduct electrical signals with precise timing so that contraction is synchronized. Cardiac arrhythmias show how catastrophic it can be when electrical coupling is wrong even if the anatomy is right. Here, bioelectricity is not merely a developmental cue; it is the ongoing operational language of the organ.

That brings us to biohybrid living robotics, a field broader than xenobots. Many groups now build systems that combine living tissues with synthetic structures: muscle rings that contract around flexible frames, neuron cultures that interface with electrodes, engineered tissues that serve as sensors because they respond to chemicals or stretch. Reviews in the mid-2020s emphasized practical constraints that tend to be ignored in popular retellings: reproducibility, durability, controllability, and biosafety. Living components can be powerful actuators and sensors, but they vary, they age, and they respond to environments in ways that can surprise engineers. Control is often indirect: through light, chemicals, temperature, or electrical stimulation. The dream is a device that is biodegradable, self-repairing, and compatible with the body. The reality is a long list of design headaches.

Those headaches, however, are exactly what makes the bioelectric approach compelling. If you can guide a tissue with signals it already uses—and if those signals can be tuned to stabilize desired patterns—then control may become less about external forcing and more about internal cooperation. There is a difference between driving a horse by dragging it and guiding it by reins that connect to its own steering. Bioelectric cues, in the best-case scenario, are closer to reins: they exploit endogenous communication.

Ethics and governance enter not as an afterthought but as part of the engineering reality. A designed living system can blur categories that matter for oversight: organism versus device, therapy versus enhancement, laboratory tool versus environmental agent. A traditional implant is regulated as a device; a cell therapy is regulated as a biological product. A computer-designed organism assembled from unedited cells sits in between. Even when it is non-reproductive outside controlled conditions, it raises questions about containment, consent (if derived from human cells in future), and purpose.

The most ethically charged issues often arise from misunderstandings of agency. When a living construct behaves adaptively, people can attribute intention where there is none, or dismiss genuine goal-directedness as mere mechanism. The more useful stance is to treat living systems as agents of a limited kind: entities with internal dynamics that pursue stability, repair, and sometimes exploration. They are not moral agents, but they are not inert objects either. That in-between status matters when deciding how to test, release, or deploy such systems.

There is also a quieter ethical issue: hype. It is easy to leap from Xenobots to fantasies of programmable organs on demand. The responsible view respects the gap between a proof-of-principle and a therapy. A millimeter-scale construct moving on a dish does not solve the vascularization problem of a human kidney. Kinematic replication in a laboratory setting does not imply uncontrolled self-propagation in nature. Bioelectric interventions in amphibian regeneration do not translate automatically to human tissues, which differ in scale, immune context, and regenerative capacity. Yet dismissing the work because it is early would be equally mistaken. The right question is not "does this already cure disease?" but "does this reveal a control handle that medicine has not fully used?"

Concrete examples make that question less abstract. Chronic wounds, such as diabetic ulcers, persist partly because the tissue fails to coordinate closure and remodeling. Electrical stimulation therapies have been explored for wound healing, and endogenous electric fields at wounds are a long-studied phenomenon. If tissue-level electrical cues help coordinate cell migration and boundary formation, then "writing" bioelectric states could become a way to push a wound out of a non-healing attractor into a healing one. That is living architecture at the scale of repair rather than replacement.

Another example sits at the edge of oncology. If cancer can be framed partly as a breakdown in tissue-level communication—including bioelectric coupling—then interventions that restore normal network states might complement genetic and chemical approaches. This is not a claim that cancer is "just electrical," but a recognition that multicellular order involves more than gene mutations. A tumor is a community of cells that has renegotiated its relationship with the body. Restoring the body's control over that community may require signals that operate at the same scale as the community itself.

The promise of synthetic morphology is therefore not a single gadget or a single treatment. It is a shift in how biological construction is imagined. Instead of treating cells as passive bricks that must be arranged into place, it treats them as active builders that can be guided by specifying goals and constraints. The earlier Xenobot work showed that changing geometry can unlock new behavior from an unchanged genome. The broader vision is that changing physiological state—including bioelectric patterns—might unlock new building behavior inside the body, allowing tissues to assemble, repair, and maintain structures that are currently beyond our manufacturing metaphors.

An organ is not a statue. It is a maintained pattern. To build one, the crucial task is not only to get the shape right on day one, but to ensure that, on day one hundred, the tissue still knows what it is supposed to be. Bioelectric networks, coupled through gap junctions and expressed through membrane potentials, offer a plausible substrate for that kind of memory: not memory as a stored sentence, but memory as a stable physiological state that the tissue returns to after disturbance. If living architecture can be engineered at all, it will likely be engineered by learning how to set and reset those states so that the cells, doing what cells naturally do, keep building the right thing.

Chapter 8

Medicine's Third Era

Medicine has evolved from the mechanics of surgery and the chemistry of pharmacology to a new frontier where the body is treated as a computational network. This chapter examines the transition from hardware-focused interventions to software-based therapies that reprogram the bioelectric states governing health, regeneration, and disease. It envisions a future where correcting a birth defect or regenerating a limb is a matter of debugging the biological code rather than micromanaging cellular machinery.

8.1 The Ion Channel Pharmacy

Modern medicine likes to tell a particular kind of story about control. Find the faulty gene, replace it. Find the misbehaving protein, block it. The body becomes a machine with a broken part, and the physician becomes an engineer of parts.

Bioelectric medicine keeps the engineering instinct but changes the unit of repair. Instead of asking only, "Which molecule is wrong?" it asks, "Which *state* is wrong?" A state is not a thing you can hold in your hand. It is an organized pattern in time: a voltage difference across a membrane, a flow of ions through a tissue, a coupling between cells that allows a group to behave as one electrical community. These patterns are invisible to our everyday senses, yet they are as real as the pressure inside a tire or the magnetic field around a compass.

The practical surprise is that medicine has been manipulating these states for a long time. It just has not always described what it was doing in those terms. Before "electroceutical" became a fashionable word, clinicians were already prescribing drugs that alter membrane voltage and excitability: anesthetics that quiet pain by dampening sodium channels, anti-arrhythmic drugs that tame cardiac rhythms by reshaping ionic currents, anticonvulsants that stabilize excitable neurons. In many cases, the drug is not "fixing" a particular gene. It is nudging a living circuit into a different regime of behavior.

That reframing matters for morphogenesis. When the target is not a symptom like pain or a rhythm like atrial fibrillation, but a *shape*—a limb, an organ boundary, a wound that heals without scarring—the object of therapy becomes the body's own pattern-control system. If tissues carry electrical "instructions" about what they should become, then an intervention that changes voltage patterns can behave less like a wrench and more like a software patch: a temporary change in control logic that allows the organism to compute its way back toward a correct anatomy.

A helpful way to hold the contrast in mind is to imagine two kinds of clinical power. One rewrites the hardware. The other debugs the software.

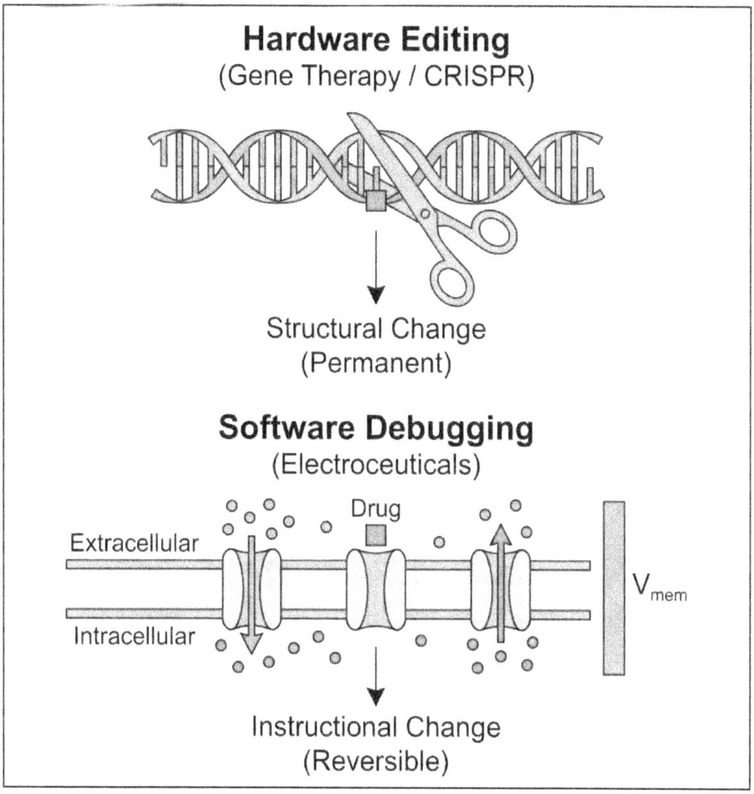

The picture is, of course, an analogy, and analogies can mislead if taken literally. "Software" in a tissue is not lines of code stored in silicon; it is a set of physical relationships that make some collective behaviors stable and others unstable. But the analogy earns its keep because it clarifies a therapeutic strategy that is easy to miss. If a disease, a birth defect, or a failure to regenerate is partly a failure of tissue-level coordination, then the intervention might not need to rebuild the parts. It might only need to push the system out of a bad attractor state and into a good one.

Ion channels are ideal levers for this kind of push. An ion channel is a protein doorway through the cell membrane that lets particular charged atoms (ions) pass. Some channels open in response to voltage; some respond to chemicals; some to stretch or temperature. Together with pumps that actively move ions against gradients, they set the cell's membrane potential, often written as V_{mem}: the voltage difference between inside and outside.

A crucial point for non-specialists is that V_{mem} is not just a detail for neurons and heart muscle. Almost every cell maintains a membrane potential. The values differ, the dynamics differ, and most cells do not fire dramatic spikes like neurons, but the basic electrical fact is universal: life runs on charge separation. Once a tissue is made of many electrically active cells connected by gap junctions (tiny bridges that let ions and small molecules pass from cell to cell), the tissue can hold patterns of voltage the way a crowd can hold a pattern of motion. Not every pattern means something, but some patterns are read by the organism as instructions: where to grow, where to stop, where to differentiate, where to maintain a boundary.

This is where the "pharmacy" enters. There is no need to invent a brand-new class of molecules to affect voltage. Pharmacology has been building an ion-channel toolkit for decades, because excitability is such a central feature of pain, seizures, arrhythmia, and many psychiatric symptoms. A "channel blocker" is a drug that reduces current through a channel. A "channel opener" increases it. Some drugs change how readily a channel opens; some change how quickly it inactivates; some alter trafficking, the process by which channels are inserted into or removed from the membrane. From the viewpoint of bioelectric patterning, these are not merely symptom suppressors. They are knobs on the tissue's electrical state.

The historical anecdote that best captures how old and foundational this is begins, appropriately, with a nerve. In the early 1950s, Alan Hodgkin and Andrew Huxley used the giant axon of the squid as a living electrical cable, measuring currents with exquisite care and building a mathematical description of how voltage-gated sodium and potassium conductances generate the action potential. Their work was basic science, motivated by curiosity about how nerves signal. Yet it also set the stage for a clinical idea that now feels almost mundane: if a brief change in membrane permeability can create a nerve impulse, then a drug that modifies those permeabilities can alter sensation, movement, and pain.

Local anesthetics such as lidocaine illustrate the principle in a tangible way. A dentist injects a small amount near a nerve. The patient stays awake, the tissue remains alive, but the pain signal does not propagate. In mechanistic terms, the drug reduces the availability of voltage-gated sodium channels, so the nerve cannot sustain the electrical spike needed for long-distance communication. The procedure feels like "turning off" an area, but what is really happening is a shift in an electrical state space. The nerve is still a nerve; it is simply held in a regime where certain behaviors are no longer available.

Electroceuticals, in the bioelectric morphogenesis sense, extend that logic beyond nerves. The target is not only whether a neuron fires, but how a group of cells holds a pattern of polarization over hours or days, and how that pattern biases gene expression, cytoskeletal organization, and tissue mechanics. The intervention may be chemical (a drug), but the therapeutic language becomes electrical: depolarize this region, hyperpolarize that boundary, restore a voltage gradient that has collapsed.

At first, it can sound like a semantic trick. If a drug binds a protein, how is this "informational" rather than "molecular"? The key is that many different molecular perturbations can lead to the same bioelectric state, and the tissue often responds to the state more than to the particular cause. This is one reason state-based therapies can be robust. If a tissue "wants" to reach a certain target morphology, and it has multiple molecular routes to get there, then restoring the right electrical constraints may allow the tissue to recruit its own pathways.

Consider an everyday example from a different corner of channel pharmacology: cystic fibrosis. The cystic fibrosis transmembrane conductance regulator (CFTR) is an ion channel that conducts chloride and bicarbonate, helping to control salt and water movement across epithelial surfaces. Certain mutations lead to thick mucus in the lungs and digestive tract, chronic infections, and severe morbidity. The newer CFTR modulator drugs are often described as precision medicine because they are tailored to the mutant protein: some act as "potentiators" that increase channel opening; others act as "correctors" that help the protein fold and reach the membrane. The clinical goal is straightforward—restore ion transport and hydration. But the deeper lesson is that an ion channel's function is not simply on or off. It can be tuned in multiple ways, and those tunings can shift a whole tissue's physiology.

In bioelectric morphogenesis, one imagines a parallel but broader aim: tune channels not only to restore a missing transport function, but to restore a pattern of voltages and couplings that tissues use to coordinate growth. A wound edge, for example, is not just a ragged line of damaged cells; it is a boundary where electrical properties change abruptly. Injury often triggers currents and voltage shifts. Some of those shifts support healing; others may lock tissue into a fibrotic, scar-forming program. If the electrical signals at the margin are part of the decision-making process, then a drug that changes channel behavior at that margin could, in principle, change the decision.

The attraction of using existing approved drugs is obvious, and it is not merely economic. Safety matters in an unusually direct way when the target is electrical state. The heart is a bioelectric organ par excellence, and its rhythms can be disrupted by many compounds. The brain is another, with its own sensitivities. A therapy that aims to shift V_{mem} in a limb bud or a healing skin patch must do so without nudging cardiac or neuronal excitability into danger. Starting with drugs that have known pharmacokinetics and known side-effect profiles is a way of narrowing the unknowns. It does not make the therapy safe by default—dose, delivery, and context are everything—but it provides a map of hazards drawn from millions of patient-years.

Repurposing also forces a change in imagination. Drug labels reflect the history of medical categories: "antidepressant," "antihypertensive," "antiarrhythmic." Those categories are organized by symptoms and organs. Ion channels are organized by physics and protein families. A sodium channel subtype in a pain-sensing neuron may share ancestry and structural motifs with a subtype in the heart. A potassium channel in an immune cell may resemble one in a developing tissue. Once the target becomes a bioelectric pattern rather than a named disease, old boundaries soften. A drug that was developed for seizures might become a tool for regeneration research, not because regeneration is "like epilepsy," but because both involve controlling membrane excitability and collective electrical dynamics.

A concrete example from developmental bioelectricity makes the point without requiring a leap into speculative human therapies. In amphibian embryos, subtle shifts in membrane potential have been shown to influence patterning decisions: where certain tissues form, how boundaries are respected, and how cells interpret chemical signals. In that experimental context, researchers can use channel openers and blockers as *actuators*—ways to impose a desired electrical condition and then observe downstream anatomy. The drug is not "a frog medicine." It is a voltage-control instrument applied to a living system that is, at that stage, busy building itself.

It is tempting to imagine this as a kind of remote control: press the hyperpolarize button, grow an organ. Real biology is messier. Electrical state interacts with chemical gradients, mechanical forces, and gene regulatory networks. If there is a "bioelectric code," it is not a single variable; it is a set of constraints and conversations among cells. Still, pharmacology has an advantage over many other forms of intervention: it can be graded, combinatorial, and reversible. A surgeon's cut is binary. A gene edit is, in many cases, permanent. A drug can be titrated, stopped, combined, and localized. The clinician can search for a regime rather than a single irreversible switch.

The idea of a "cocktail" can sound crude, like guessing. But consider how the nervous system itself achieves specificity. Neurons do not rely on one channel type. They rely on an orchestra: sodium channels for the upstroke of a spike, potassium channels for repolarization, calcium channels for release of neurotransmitter, chloride channels for inhibition, pumps to restore gradients. Tissue-level bioelectric patterns are likely to be similarly combinatorial. If so, the most effective electroceuticals may not be silver bullets but carefully tuned mixtures that shift a network of conductances into a desired steady state.

There is another reason the "software patch" metaphor is clinically seductive: it offers a way to treat disorders where the genome is not "wrong" in any simple sense. Many congenital malformations and many failures of healing arise from small perturbations early in development—a transient insult, a disrupted signaling pathway, an inflammatory event—that send the system down a different path. The genome remains unchanged; the outcome is altered. A state-based intervention is conceptually matched to a state-based problem. It does not need to rewrite the DNA if the trouble is that the tissue has adopted a self-stabilizing but undesirable pattern.

Even in cancer biology, where mutation is a central theme, electrical state has become harder to ignore. Cancer cells often show altered membrane potentials and altered coupling to neighbors. It is not that voltage "causes" cancer in a simplistic sense, but that the electrical integration of a tissue is part of how it enforces cooperation. A healthy tissue is a society of cells with rules. Gap junctions help enforce those rules by allowing cells to share signals, synchronize behaviors, and sense when a neighbor is out of step. Disruptions in coupling and polarization can loosen that enforcement. From that viewpoint, some cancers may be understood partly as failures of tissue-level governance, with bioelectric patterns acting as one medium of governance.

If that sounds abstract, return to a more familiar clinical experience: the way certain psychiatric drugs take weeks to exert their full effects, even though they bind receptors quickly. The brain is not a bag of independent molecules; it is a network with attractor states. A small molecular nudge can, over time, lead to network rewiring and state transitions. Tissue morphogenesis is also a network problem, unfolding over time. Electroceuticals aim to exploit that same leverage: small, controlled perturbations that allow the system to reorganize itself.

The most practical objection is also the most important: if ion channel drugs are so powerful, why are they not already used to grow back limbs or erase scars? The honest answer is that changing bioelectric state is easy; changing it *precisely* is hard. The channels targeted by a pill are distributed across the body. The same class of channel may do something helpful in a wound edge and something dangerous in cardiac conduction tissue. Dose that is too low may do nothing; dose that is too high may cause toxicity. Timing matters: a transient electrical cue at a critical developmental window can have large downstream effects, while the same cue later may be irrelevant. Delivery matters: a local tissue needs a local intervention.

That is why the "ion channel pharmacy" is best seen as a foundation rather than a finished technology. It is a stockroom of already-invented actuators for changing electrical state. Some are blunt, some surprisingly specific, and the trend in channel pharmacology has been toward greater subtype selectivity and more sophisticated mechanisms, including drugs that stabilize particular conformations of a channel protein. Those details may sound like specialist concerns, but they matter for morphogenesis because they determine whether one can shift voltage in one anatomical region without causing unwanted changes elsewhere.

There is also a quieter, human reason this approach is attractive. Gene editing raises understandable fears: permanence, inheritance, unintended consequences that cannot be recalled. State-based therapy suggests a different relationship to risk. If a treatment acts primarily by shifting a physiological pattern, one can imagine cautious protocols: apply locally, monitor continuously, withdraw if the trajectory is wrong. The promise is not that the therapy is trivial, but that it belongs to a family of interventions medicine already knows how to manage: adjustable ones.

The body's shapes are not sculpted only by what genes encode, but by what cells collectively decide to build and maintain. Ion channels sit at a strategic chokepoint in that decision-making: they translate chemistry into electricity, electricity into cellular behavior, and cellular behavior into tissue-level form. A pharmacy that can tune those channels is more than a catalog of symptom treatments. It is, potentially, a set of keys to the electrical control layer that life uses to remember what it is supposed to look like.

8.2 Crossing the Mammalian Gap

A flatworm can lose its head and regrow a new one, complete with a brain. A frog tadpole can rebuild parts of its tail. These feats are not delicate miracles; they are repeatable, robust, and, in the right animals, almost routine. People hear this and instinctively ask the obvious question: if nature already knows how to do it, why is adult human regeneration so limited?

The discouraging answer is not that mammals lack the right genes. Much of the molecular toolkit for growth is conserved across animals. The hopeful answer is also not that mammals secretly regenerate whole limbs and we have simply overlooked it. The truth lies in a more interesting middle ground: mammals *do* regenerate in certain contexts, but the conditions are tightly constrained, and the default response to major injury is not rebuilding but stabilizing. A mammalian wound is treated as a breach to be sealed quickly, even if that means laying down scar tissue that is mechanically useful but anatomically crude.

Crossing the mammalian gap means learning to flip that default. Bioelectric interventions are appealing because they aim at the control layer that decides what kind of healing is underway. Yet the gap is real, for reasons that are as practical as they are philosophical. A frog embryo in a dish is a forgiving experimental universe: transparent, accessible, and tolerant of big perturbations. A human body is opaque, guarded, electrically noisy, and ethically non-negotiable. A voltage gradient that can be painted across a tadpole with a few molecules becomes, in a person, an intervention that must avoid disturbing heart rhythms, brain function, and the countless electrical conversations occurring in muscles, glands, immune cells, and epithelia.

The first step toward translation is therefore a change in expectations. The goal is not to take a spectacular trick from a "regenerative champion" and simply scale it up. It is to find mammalian footholds: places where regeneration already happens, even modestly, and ask what keeps the door open there.

A striking foothold sits at the end of your finger.

Children and, to a lesser extent, adults can regrow the very tip of a digit if the amputation is distal enough—closer to the nail than to the knuckle. The regrown tissue is not a perfect replica, and the conditions are fussy, but the phenomenon is real: bone, skin, and nail structures can reconstitute rather than merely scar over. Biologists describe the formation of a blastema-like structure, a temporary assembly of cells that proliferates and coordinates rebuilding, reminiscent of the classic blastema of salamander limb regeneration. The nail epithelium and peripheral nerves appear to matter, not as decorative extras but as sources of signals and organization. Modern single-cell methods have begun to map which cell states appear and when, turning what was once a descriptive curiosity into a programmable target.

The fingertip case does two things at once. It shows that "mammals cannot regenerate" is simply false. It also shows why translation is hard: regeneration is not an on/off ability that one either has or lacks. It is a state that depends on location, age, immune tone, nerve input, mechanical context, and, very plausibly, bioelectric context.

If regeneration is a state, then it has to be entered. And entering a state often requires crossing a threshold.

A historical anecdote helps here, not from regeneration research but from cardiology. In the early days of studying arrhythmias, it became clear that the heart could be pushed from one rhythm to another by a well-timed electrical shock. The shock does not "repair" any molecular defect. It resets a collective electrical pattern. The idea behind defibrillation is not mystical: it is dynamical systems thinking applied to tissue. You interrupt a dangerous attractor and allow the system to fall back into a stable rhythm.

Bioelectric regenerative medicine hopes for a gentler version of the same logic: not a dramatic reset, but a controlled nudge of tissue-level voltage patterns so that wound healing falls into a regenerative attractor rather than a fibrotic one. The difficulty is that the heart is a specialized organ designed to be synchronized and reset. A wound edge is a heterogeneous landscape of dying cells, immune infiltrates, stressed epithelia, exposed extracellular matrix, and disrupted nerves. Its electrical behavior is less like a metronome and more like a city after an earthquake.

The second foothold comes from an animal that looks, at first glance, like an ordinary mouse. The spiny mouse, *Acomys*, has earned attention because it can regenerate complex skin and ear tissue with surprisingly little scarring, restoring hair follicles and other structures that typical laboratory mice do not. This is not the regrowth of a limb, but it is a meaningful departure from the mammalian default. It suggests that the mammalian gap is not a wall. It is a set of dials that evolution has turned.

One of those dials appears to be inflammation. Another is tissue mechanics. A third is the persistence of pro-growth signaling pathways. Work on *Acomys* has highlighted the importance of sustained ERK activity: a signaling route often associated with growth and repair. The striking lesson is not that ERK is “the regeneration molecule.” It is that regenerative healing may require signals to remain on longer, in the right places, without tipping into uncontrolled growth. Mammals often respond to injury by rapidly engaging programs that close the wound and then shut down. The body prefers a quick, safe patch over an ambitious rebuild.

Bioelectric control could, in principle, provide a way to keep the right programs active without simply bathing the tissue in growth factors. Voltage patterns are not chemicals that diffuse indiscriminately; they can, with the right tools, be spatially organized. They can also integrate with the immune response. Immune cells themselves are electrically active, with membrane potentials and ion channels that shape activation, migration, and cytokine release. A wound's electrical environment is part of what immune cells sense and alter. In that sense, bioelectric interventions might influence regeneration not only by speaking to stem and progenitor cells, but by tuning the inflammatory conversation that decides whether scar or structure wins.

Still, the most stubborn translational problems are engineering problems: specificity, delivery, and measurement.

Specificity comes first because the mammalian body is full of excitable tissues. A drug or device that modulates ion channels can have off-target effects that are not subtle. Slightly changing cardiac repolarization can provoke dangerous arrhythmias. Slightly changing neuronal excitability can trigger seizures or cognitive side effects. Even if an intervention is intended for a wound on the skin, systemic exposure can spread the effect.

This is where the glamour of "electroceuticals" can be misleading. A pill is not automatically an electrical therapy in the morphogenetic sense. If it shifts V_{mem} everywhere, it is more like changing the voltage of an entire city when you only wanted to fix one streetlamp. Translating bioelectric control therefore pushes medicine toward localized or gated delivery: apply the actuator only where and when it is needed, and keep it from affecting electrically sensitive organs.

Delivery is the physical counterpart to that idea. How do you get a channel modulator, a voltage-sensitive actuator, or an electrical stimulus to the right cells at the right time?

In simple animals and embryos, the answer can be "put it in the water," which is both wonderfully convenient and completely unacceptable as a clinical strategy. In mammals, delivery becomes a design space that includes wound dressings, injectable hydrogels, nanoparticles, liposomes, and implantable devices. It also includes clever gating methods: drugs that remain inactive until illuminated, for example, or materials that release their cargo only under certain pH or enzymatic conditions typical of wounded tissue.

The wound itself offers an opportunity, because it is accessible. A skin injury can be dressed; a fingertip can be covered; a surgical bed can be irrigated. The clinical imagination shifts from systemic therapy to microenvironment engineering: treat the wound as a temporary organ with its own physics and chemistry, and build a niche that favors regeneration.

A concrete example of this philosophy is the BioDome device developed around 2010: a sleeve-like system for mouse digit injuries designed to control the wound environment while allowing combinations of biochemical cues and biophysical stimulation, including electrical stimulation. The point was not that a single component is magical. The point was methodological: treat regeneration as a systems problem and build an apparatus that can deliver multiple controlled inputs. If a digit tip can regenerate under certain conditions, then a device that standardizes those conditions makes the biology legible and the intervention reproducible.

Measurement is the third engineering pillar, and it is often underestimated. If the target is an electrical state, clinicians need to know whether that state has been achieved. Yet most clinical tools measure anatomy (imaging), chemistry (blood tests), or bulk electrical activity (ECG, EEG). The relevant bioelectric patterns for morphogenesis may be slower, subtler, and more spatially structured: gradients across a tissue, local depolarized "islands," changes in coupling between cell groups.

In research, voltage-sensitive dyes can reveal these patterns in striking colors. But dyes are not trivial to use in humans, and they raise safety and regulatory questions. Noninvasive electrical impedance measurements can provide indirect information about tissue state, but spatial resolution is limited. Implantable sensors can read local potentials, but implants bring their own risks: infection, fibrosis around electrodes, and long-term stability. The measurement problem is not a mere technical detail; it shapes what kinds of therapies can be tested. If you cannot reliably read the variable you claim to control, you cannot run clean trials, and you cannot convince regulators.

Even when measurement is possible, the translation problem has a conceptual trap: confusing "state normalization" with "symptom improvement." A therapy that makes a wound look better at two weeks may still fail to restore function at six months. A therapy that reduces inflammation might reduce pain but increase infection risk. Regeneration is a long game, and bioelectric interventions may have time-dependent effects: what helps early might harm later.

That time-dependence makes targeting more complex. Many developmental systems are exquisitely sensitive to timing: a brief perturbation during a narrow window can have large downstream consequences. The adult wound may have analogous windows. Early after injury, cells decide whether to proliferate, migrate, differentiate, or secrete matrix; immune cells decide whether to resolve or sustain inflammation; nerves sprout or retract. Bioelectric patterns likely play different roles at different times. Delivering an intervention on day one may not resemble delivering it on day seven, even if the drug is identical. Translation therefore requires a temporal map, not just a molecular map.

Regulation adds another layer of difficulty, because state-based therapies fit awkwardly into categories designed for molecule-based mechanisms. Traditional pharmacology likes a linear story: drug binds target, target changes pathway, pathway changes symptom. Bioelectric medicine often tells a more circular story: intervention shifts tissue state, tissue state changes signaling networks, networks stabilize a new state. That can look suspiciously like hand-waving unless it is backed by rigorous biomarkers and mechanistic anchors. Regulators do not require a complete philosophical theory of morphogenesis, but they do require evidence of safety, reproducibility, and clinically meaningful benefit.

The most sobering constraint is also the most human: success in regenerative bioelectricity is easiest to demonstrate in organisms that tolerate radical mistakes. A planarian can be cut into pieces and still live. An embryo can be perturbed and still yield a viable organism, often with surprising robustness. Humans are less forgiving. The cost of error in a developing fetus or a healing adult organ is not an odd-looking animal in a dish; it is disability or death. That reality does not make translation impossible, but it raises the bar for precision.

Precision, in practice, may come from combining modalities rather than betting on one. Chemical electroceuticals can adjust ion channels. Devices can deliver spatially confined electrical stimulation. Materials can shape the wound microenvironment mechanically and chemically. Computational models can suggest which combinations are likely to push tissue toward a desired electrical pattern without destabilizing excitable organs. The emerging vision is not a single miracle switch but a controlled "regenerative niche," engineered in space and time.

The most useful mental model for crossing the mammalian gap is therefore not copying what salamanders do, but asking what mammals are already willing to do when the constraints are just right. The fingertip regenerates at the boundary between nail and bone; the spiny mouse regenerates skin structures that ordinary mice scar over. These are not trivial exceptions. They are demonstrations that mammalian tissues can enter regenerative states, and that evolution has not erased the circuitry required to do it. It has placed that circuitry behind locks.

Bioelectricity may provide one set of keys, but keys only matter if you can get them into the right lock without setting off alarms elsewhere. The practical work of translation is full of such alarms: arrhythmia risk, seizure risk, immune disruption, tumor risk, infection risk. Those are not reasons to abandon the project. They are the defining constraints that make it a real medical science rather than a fairy tale.

A mammal heals not only with cells and molecules but with priorities: stop bleeding, prevent infection, restore barrier, preserve function. Scar tissue is a blunt instrument, but it is fast and, in evolutionary terms, often good enough. Any regenerative therapy has to persuade the body that rebuilding is safer than patching. Bioelectric interventions, by acting on the signals that coordinate collective decisions, offer a plausible way to deliver that persuasion—not as an argument in words, but as a carefully shaped pattern that the tissue can recognize as a cue to organize rather than merely to seal.

8.3 The Ethics of Morphological Freedom

Imagine a clinic that does not promise a pill for pain or a scalpel for a tumor, but a dial for form. The clinician does not merely remove what is there; they help a body decide what it should become. A scar can be persuaded to remodel into supple skin. A malformed jaw can be guided, slowly, toward symmetry. A damaged nerve can be coaxed to reconnect not by stitching, but by restoring a pattern of signals that says *this tissue belongs together*. The temptation is to call such powers futuristic, but the ethical questions they raise are already here, because medicine has long been in the business of shaping bodies. What changes is the ease and scope of the shaping, and the identity of the shaper.

"Morphological freedom" is a phrase meant to sound like a civil liberty. In its most straightforward meaning, it is the claim that a person should have the right to keep their body as it is, or to modify it, on their own terms, through informed and consensual access to technologies that make such choices possible. The definition contains two protections. One is a right *to modify*. The other, equally important, is a right *to refuse modification*. A world that celebrates bodily redesign can become a world that punishes the unredesigned.

The moral appeal is easy to feel because bodies are not abstract property. They are the place where pain happens, where pleasure happens, where dignity is won or denied. A person born with a craniofacial condition may spend childhood being stared at, questioned, treated as a lesson for strangers' children. A person with a spinal cord injury may spend adulthood navigating buildings and institutions designed as if wheelchairs were a rare inconvenience rather than a common mode of life. The shape of a body, and the assumptions others make about that shape, can determine education, employment, romance, safety, and self-respect.

If bioelectric control makes it easier to alter growth and repair, it will not be experienced as a neutral technical advance. It will land in a world with existing inequalities and existing norms about what bodies *ought* to look like. The ethics, then, begin with a deceptively simple question: who gets to decide what counts as a defect, a difference, a disability, an identity, or an acceptable variation?

A concrete historical example shows how quickly therapeutic reshaping can slide into social pressure. In the 1920s and 1930s, plastic surgery expanded rapidly, partly driven by the aftermath of World War I, when surgeons confronted terrible facial injuries and developed techniques of reconstruction. For many patients, reconstruction was an act of restoration: relief from disfigurement, a chance to re-enter social life without constant scrutiny. Yet the same surgical tools also became instruments for elective "normalization," shaped by cultural ideals of beauty and racialized concepts of facial features. What began as repair did not remain purely repair. The technology did not dictate the values; the society did.

Bioelectric morphogenesis technologies would be even more value-sensitive, because they would not only sculpt surfaces but could, in principle, influence the deep organization of tissues. A therapy that encourages regeneration over scarring is easy to celebrate. Few people feel an attachment to fibrosis. But once the same logic can "optimize" height, muscle distribution, facial symmetry, or even the architecture of sensory organs, the line between healing and enhancement stops looking like a line and starts looking like a gradient.

Some gradients are ethically manageable. Orthodontics already moves bones. Growth hormone therapy already alters stature. Cochlear implants already reshape sensory experience, and their use has generated profound debates within Deaf communities about identity, culture, and the framing of deafness as a defect to be corrected. The cochlear implant debate matters here because it highlights a recurring ethical pattern: a technology can be individually beneficial while also sending a collective message that a way of being is inferior. Even when parents choose implants out of love, the social meaning of the choice is not purely private.

Bioelectric interventions could intensify this tension by making bodily change less invasive and more scalable. A brace and years of orthodontic adjustments are burdensome. A surgery is daunting. But a treatment that quietly modifies tissue growth through a drug regimen or wearable stimulator could be perceived as almost trivial, the way cosmetic injectables are sometimes marketed as lunchtime procedures. When an intervention feels trivial, the pressure to undergo it becomes harder to resist, because refusal can be framed as irrational or irresponsible.

This is where the "freedom" in morphological freedom becomes ethically fragile. Freedom is not only the absence of prohibition; it is also the absence of coercion, including soft coercion. Employers, insurers, schools, and even family members can apply pressure without ever issuing an explicit mandate. If a therapy reduces future medical costs, insurers may quietly incentivize it. If a modification improves productivity, employers may prefer modified bodies. If a change is marketed as "preventive," parents may feel compelled to choose it for their children, not because they are persuaded it is best, but because they fear being blamed later.

The stakes become sharper when the choice is made on behalf of someone who cannot consent. Much of pediatric medicine already operates in this space, and often rightly: infants receive lifesaving interventions they did not request. Yet morphogenetic control adds a particular twist. Some interventions could alter not merely health outcomes, but the trajectory of identity. If a technology can modify facial structure, limb form, or sensory capacities early in life, the child grows up as someone for whom that form was never an option to refuse. Ethical frameworks for pediatric decision-making rely on notions like "best interests" and the child's "right to an open future." Those notions become harder to apply when the intervention is not simply preventing death or disease, but selecting among ways of being.

A person might object: all of this sounds like the standard bioethics debate about enhancement, and perhaps it is. The distinctive feature of bioelectric morphogenesis is that it reframes what is modifiable. Traditional enhancement debates often assume discrete upgrades: a gene edit, a prosthetic implant, a drug that improves attention. Morphogenetic control suggests *form itself* could be a medium of choice.

That brings the discussion uncomfortably close to older, darker histories.

Eugenics did not begin with genetic sequencing. It began with social ranking of bodies and minds, with the belief that a "better" population could be engineered by controlling reproduction and, more broadly, by controlling who counted as normal. The atrocities of compulsory sterilization and racial hygiene policies were enabled by pseudoscience, but also by a mundane bureaucratic confidence: the state believed it could define and enforce biological standards. Any technology that expands the range of possible bodily forms must be defended against that style of governance. This is not alarmism; it is historical memory applied to a new tool.

The rights-based argument for morphological freedom is partly a response to this. It emphasizes self-ownership and bodily autonomy, and warns that centralized control of body modification technologies can become coercive. It also insists on tolerance: if bodies become more modifiable, diversity of bodies must become more acceptable, not less. Otherwise "choice" becomes a one-way street toward conformity.

Yet rights language can obscure another ethical fact: bodies are not only private projects; they are also public realities. A person's body affects others in tangible ways. Some modifications can increase risk to bystanders (for example, if an enhancement increases aggression, or if an intervention increases infection risk). Some modifications can reshape social norms in ways that harm those who did not choose them. Even benign enhancements can become collectively coercive. If an intervention improves endurance or attention, the baseline expectations of schools and workplaces can shift, making the unenhanced seem lazy or incompetent. A world where the average human runs faster is a world where the slow are judged more harshly.

Disability ethics has long insisted on a distinction that is easy to forget in a laboratory: disability is not only a property of a body; it is also a property of an environment. A wheelchair is disabling in a building without ramps, and liberating in a building designed for access. A person with atypical sensory processing is disabled in a rigid educational system, and can thrive in a flexible one. Morphological freedom, if interpreted narrowly as "fix the body," risks reinforcing the idea that the world should not have to change. It can become a polished version of an old demand: adapt to the norm, because the norm will not adapt to you.

Here a concrete, contemporary legal issue illuminates the stakes. Many people experience assistive technologies not as detachable tools but as integrated parts of the self: a prosthetic limb, a wheelchair, a communication device. Legal systems and social attitudes sometimes treat these devices as external property, separable from personhood, which can create vulnerabilities. When a mobility device is damaged, confiscated, or denied access, the harm is not like losing a suitcase; it is like being partially immobilized. The boundary between "person" and "device" is ethically porous. If bioelectric interventions begin to integrate with wearables, implants, or responsive materials that become part of everyday function, personhood will be even harder to separate from technology. Rights protections may need to evolve so that interfering with an enabling device is recognized as interfering with the person.

The prospect of "open-source" electroceuticals adds another ethical twist. One can imagine a democratizing narrative: if bioelectric control can be achieved with repurposed drugs and inexpensive devices, then bodily self-determination becomes widely accessible. The same tools, however, could enable reckless self-experimentation, black-market clinics, and a new genre of coercion within communities and families. Technology that is easy to distribute is not automatically easy to govern.

A pragmatic response is to borrow governance templates from areas that have already wrestled with the therapy-enhancement boundary and the problem of global coordination. Genome editing debates, especially after the birth of gene-edited babies was announced in 2018, forced international bodies to articulate principles: transparency, independent oversight, proportionality of risk to benefit, special caution for interventions that could affect future generations, and attention to equity. Bioelectric morphogenesis is not germline editing, but it raises parallel concerns about irreversible changes, long-term unknowns, and the temptation to race ahead in jurisdictions with weak regulation.

Oversight will have to contend with a peculiar challenge: state-based therapies can be hard to define in traditional regulatory language. A gene edit changes a sequence. A drug has an active ingredient. A bioelectric intervention may be a combination of a mild electric field, a local drug, and a smart dressing that modulates ion flow. The "active ingredient" is arguably a physiological state rather than a molecule. Regulators can adapt to that, but only if the science produces measurable markers of state and clear links to outcomes. Otherwise, the field risks becoming a playground for exaggerated claims, with patients as the experimental substrate.

Equity is not an optional add-on to these considerations; it is central. Technologies that reshape bodies can deepen inequality in two ways. First, by unequal access: those who can pay gain restorative or enhancing options that others do not. Second, by unequal standards: the bodies of the disadvantaged are more likely to be treated as problems to be corrected in order to fit hostile environments. If workplace injuries are common in low-wage labor, will regenerative technologies be used to heal workers while leaving the dangerous conditions unchanged? If a "correction" is available for a trait that attracts discrimination, will society feel less obligated to fight the discrimination?

There is also a subtler equity issue: whose bodies become the template for "normalization" algorithms and clinical standards. Medicine has a long history of treating certain populations as the default and others as deviations. If morphogenetic therapies rely on computational models of "ideal" tissue states, those models could encode bias. An algorithm trained on limited datasets might misinterpret normal variation as pathology. A voltage pattern that is healthy in one population might be flagged as abnormal in another. Ethical bioelectric medicine will require not only consent and oversight, but careful attention to the representativeness of the biological baselines it treats as targets.

The most emotionally charged question, however, is the one that returns in every enhancement debate: does freedom to enhance become an obligation to enhance?

Consider a hypothetical but plausible scenario. A bioelectric therapy can reduce scarring after major surgery, producing more flexible tissue and fewer long-term complications. Soon it becomes standard of care. Then, a related therapy can increase tendon resilience and reduce injury risk in athletes. It is offered to professionals, then to college recruits, then to ambitious teenagers. At each step, the therapy can be framed as voluntary. Yet the meaning of "voluntary" shifts when the alternative is falling behind.

The obligation-to-enhance can also creep in through moral language. A parent might be told that refusing a safe enhancement is depriving a child of opportunities. An adult might be told that declining a modification that could reduce healthcare costs is selfish. A soldier might be told that enhancement is part of the job. The ethics of morphological freedom must therefore include protections not only for access, but for refusal without penalty.

Refusal becomes especially important when the body is treated as a project of self-expression. For some people, an atypical body is not a tragedy to be corrected but a meaningful identity. A person may value the visible marks of survival: scars, differences, prosthetics, the evidence of a life lived. A world that offers easy smoothing and "optimization" risks making such bodies less legible and less socially acceptable. It also risks narrowing our moral imagination. If every difference is seen as a fixable defect, compassion can wither into impatience.

At the same time, it would be cruel to romanticize suffering or to demand that people carry burdens in the name of diversity. Ethical reflection must avoid turning disability or disfigurement into a moral performance. The point of morphological freedom is not to celebrate pain, but to allow people to decide what kinds of pain and effort they are willing to endure, and what kinds of interventions they want, without being forced by stigma or by policy.

A final ethical knot concerns responsibility. If a person modifies their body in a way that carries risk, who bears the cost when something goes wrong? Individuals? Insurers? The state? What if a modification has long-term consequences that appear years later? These questions are familiar from elective surgery and from some pharmaceutical risks, but they become more complex when modifications are distributed, incremental, and state-based. A person might undergo a sequence of "small" interventions whose interactions are poorly understood. The ethical demand for informed consent becomes harder to satisfy when the information is genuinely uncertain.

In the end, the ethics of morphological freedom are less about a single rule than about sustaining a humane stance toward bodily difference in a world of expanding control. A society can embrace regenerative medicine without turning bodies into compliance objects. It can support enhancement research without punishing the unenhanced. It can protect experimentation without abandoning people to predatory markets. These are not purely technical achievements. They are cultural achievements, made out of law, education, design, and the everyday choices of how we look at one another.

If the body plan becomes more rewritable, the deepest test will be whether the category of the "normal body" loosens rather than tightens. The most ethical future would not be one in which everyone converges on a single optimized morphology, but one in which more people can inhabit their own forms without apology, with access to restoration when they want it, and with protection from correction when they do not.

www.ingramcontent.com/pod-product-compliance
Lightning Source LLC
LaVergne TN
LVHW022211200926
848481LV00006B/1925
9798896652885